73
CRIME & DETECTION
74
RUSSIA
75
LIGHT
76
ENERGY
77
ELECTRICITY
78
FORCE & MOTION
79
80
81
TIME & SPACE
82
ASTRONOMY
83
EARTH
84
LIFE
85
EVOLUTION
86
ECOLOGY
87
HUMAN BODY
88
MEDICINE
89
TECHNOLOGY
90
ELECTRONICS
91
RENAISSANCE
92
IMPRESSIONISM
93
GOYA
94
MANET
95
MONET
96
VAN GOGH
97
WATERCOLOR
98
PERSPECTIVE
99
DANCE
100
FUTURE
101
MYTHOLOGY
102
LEONARDO & HIS TIMES
103
OLYMPICS
104
MEDIA & COMMUNICATION
105
TITANIC
106
FOOTBALL
107
HURRICANE & TORNADO
108
SOCCER
109
PRESIDENTS
110
BASEBALL
111
EPIDEMIC
112
WORLD WAR II
113
SUPER BOWL
114
CIVIL WAR
115
RESCUE
116
EVEREST
117
FIRST LADIES
118
WORLD WAR I

DORLING KINDERSLEY EYEWITNESS BOOKS

ASTRONOMY

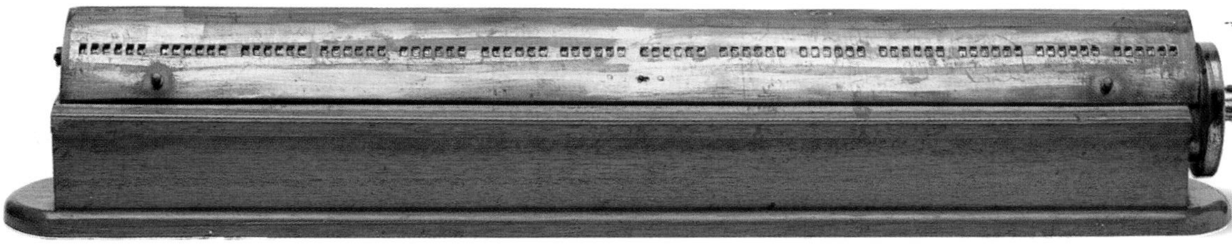

Calculator
(19th century)

Model of Stonehenge

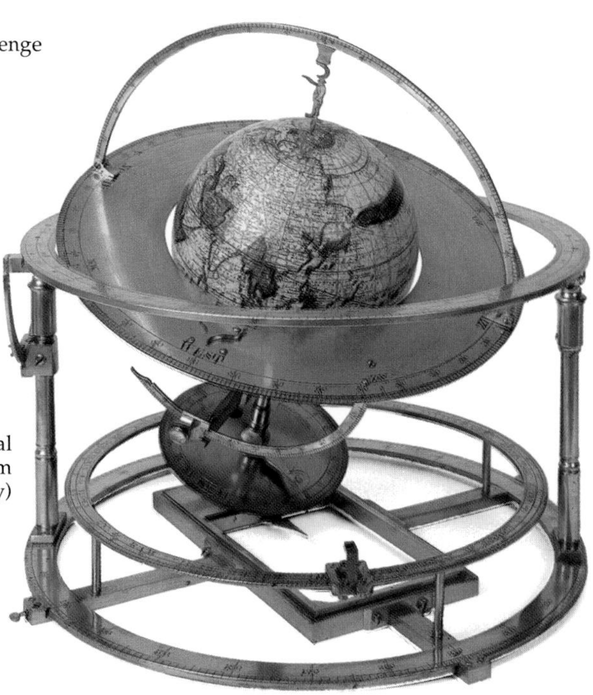

An ornamental
cosmotherium
(19th century)

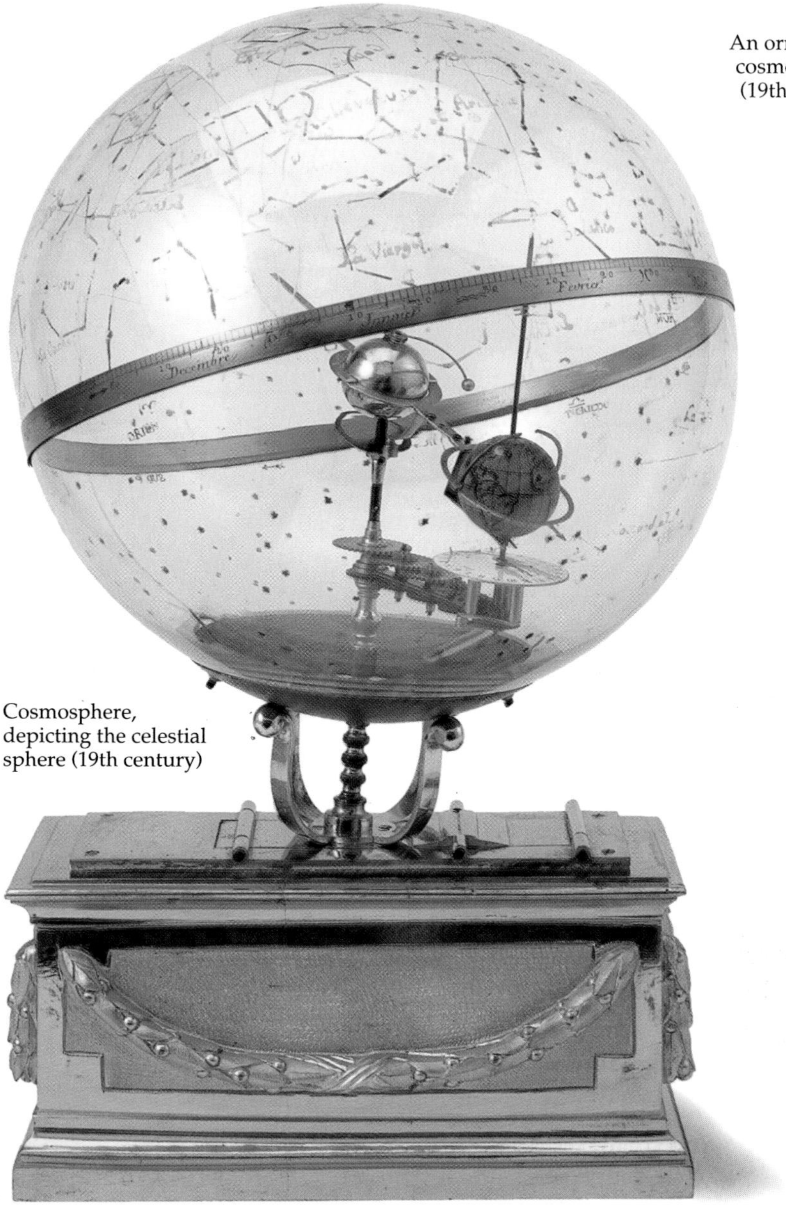

Japanese sundial
(19th century)

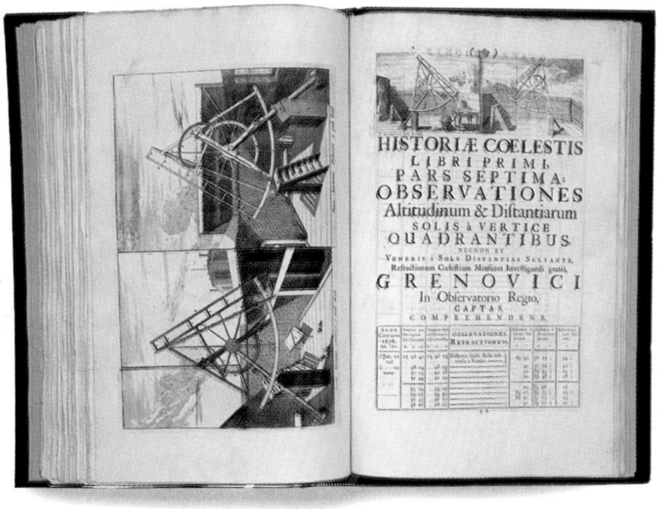

Cosmosphere,
depicting the celestial
sphere (19th century)

The star catalog of John Flamsteed (1725)

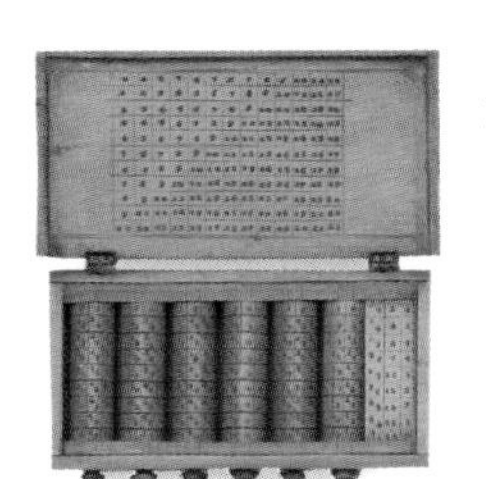

Napier's bones

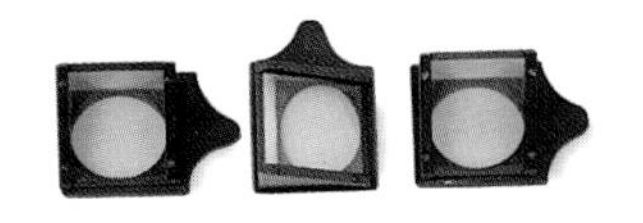

Prisms used in a 19th-century spectroscope

ASTRONOMY

Written by
KRISTEN LIPPINCOTT

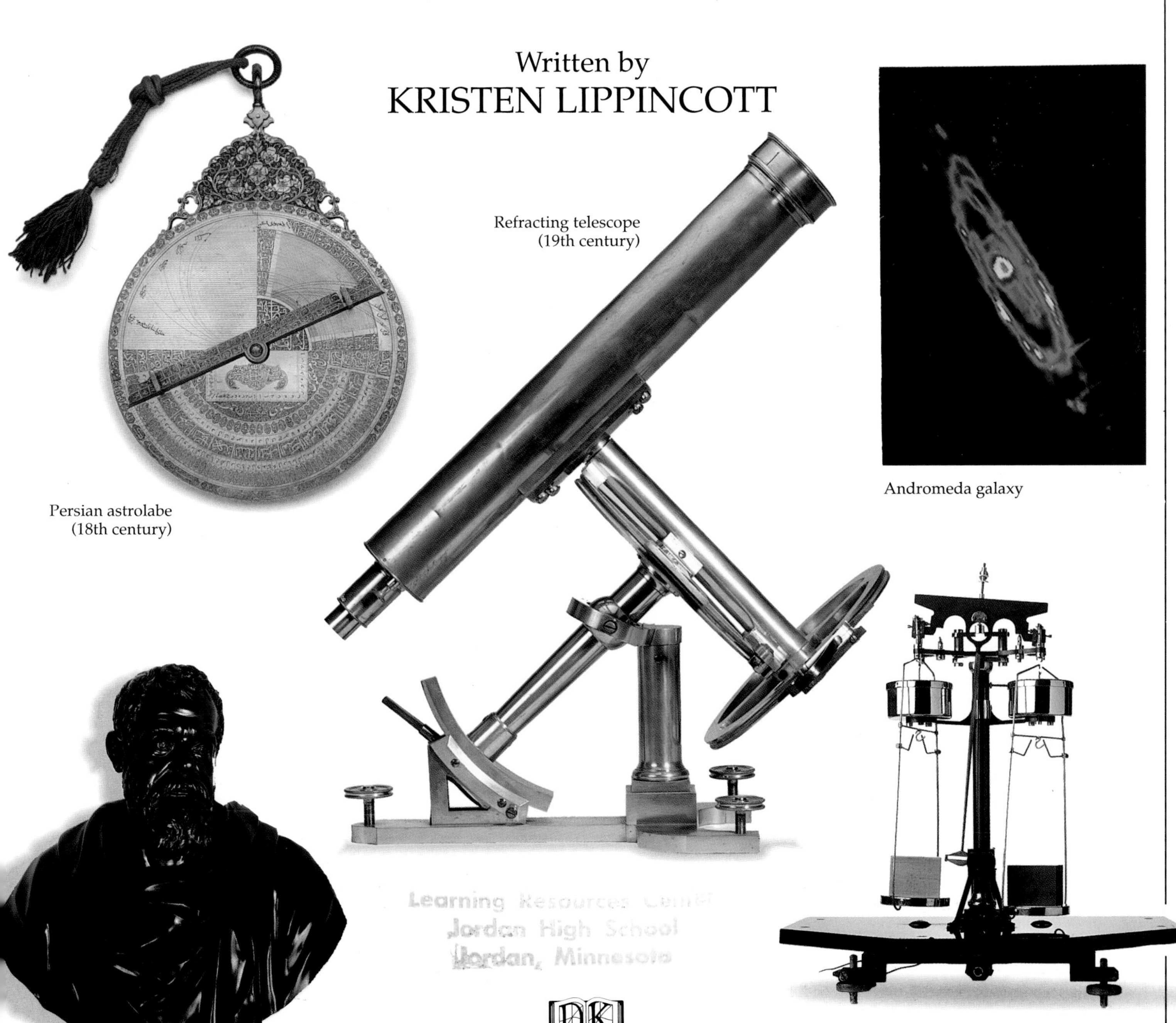

Refracting telescope (19th century)

Andromeda galaxy

Persian astrolabe (18th century)

Bust of Galileo

Beam balance to find mass

Dorling Kindersley

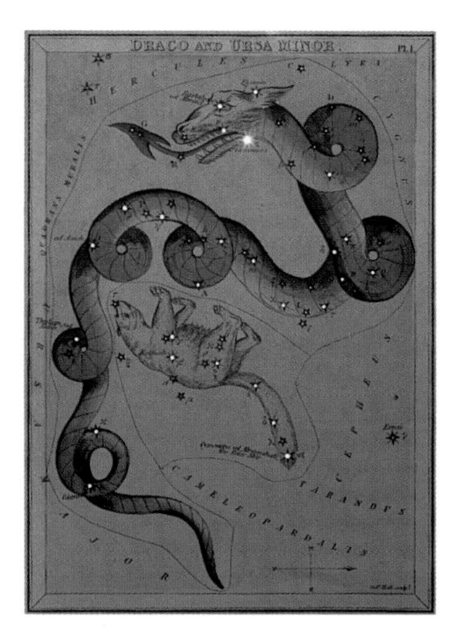

Painted constellations on a 19th-century playing card

Dorling Kindersley
LONDON, NEW YORK, AUCKLAND, DELHI, JOHANNESBURG, MUNICH, PARIS and SYDNEY

For a full catalog, visit

Project editor Charyn Jones
Art editor Ron Stobbart
Design assistant Elaine Monaghan
Production Meryl Silbert
Special photography Clive Streeter
Picture research Deborah Pownall and Becky Halls
Managing editor Josephine Buchanan
Managing art editor Lynne Brown
Editorial consultant Heather Couper
US editor Charles A. Wills
US consultant Dr William A. Gutsch

This Eyewitness ® Book has been conceived by Dorling Kindersley Limited and Editions Gallimard

First American edition, 1999

Published in the United States by
Dorling Kindersley Publishing, Inc.
95 Madison Avenue
New York, NY 10016
4 6 8 10 9 7 5 3

Library of Congress Cataloging-in-Publication Data
Lippincott, Kristen.
Astronomy / written by Kristen Lippincott.
p. cm. — (Eyewitness Books)
Includes index.
1. Astronomy—Juvenile literature.
[1. Astronomy.]
I. Title. II. Series.
QB46.L744 2000 550—dc20 94-18479
CIP
AC

ISBN 0-7894-6179-X (pb)
ISBN 0-7894-4888-2 (hc)

Color reproduction by Colourscan, Singapore
Printed in China by Toppan Printing Co. (Shenzhen) Ltd.

Compass (19th century)

Drawing an ellipse

Micrometer for use with a telescope

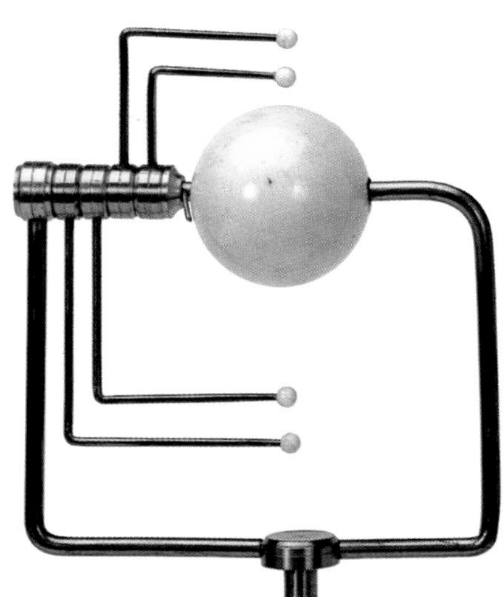

A planetarium to show the planet Uranus and its satellites

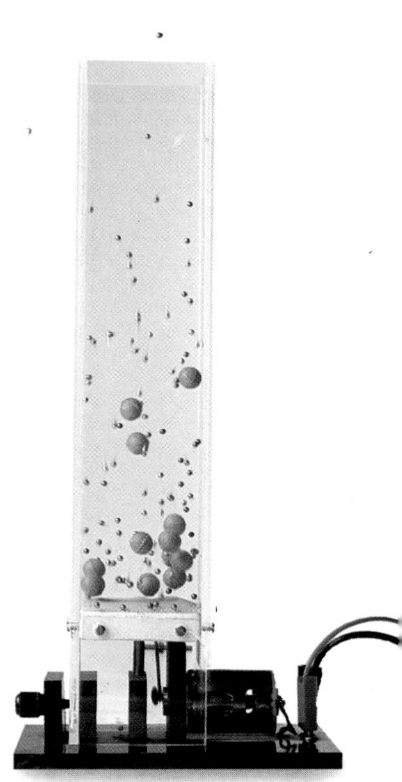

A demonstration to show how different elements behave in the Solar System

Contents

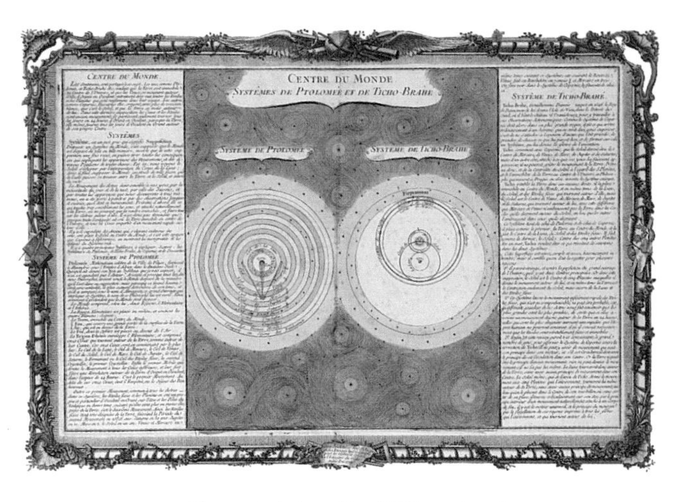
Illustrated star atlas (19th century)

The study of the heavens

THE WORD "ASTRONOMY" comes from a combination of two Greek words: *astron* meaning "star" and *nemein* meaning "to name." Even though the beginnings of astronomy go back thousands of years before the ancient Greeks began studying the stars, the science of astronomy has always been based on the same principle of "naming the stars." Many of the names come directly from the Greeks, since they were the first astronomers to make a systematic catalog of all the stars they could see. A number of early civilizations remembered the relative positions of the stars by putting together groups that seemed to make patterns in the night sky. One of these looked like a curling river, so it was called Eridanus, the great river; another looked like a hunter with a bright belt and dagger and was called Orion, the hunter (p. 61). Stars were named according to their placement inside the pattern and graded according to brightness. For example, the brightest star in the constellation Scorpius is called α *Scorpii*, because α is the first letter in the Greek alphabet. It is also called Antares, which means "the other Mars," because it shines bright red in the night sky and strongly resembles the blood-red planet, Mars (pp. 48-49).

WATCHING THE SKIES
The earliest astronomers were shepherds who watched the heavens for signs of the changing seasons. The clear nights would have given them opportunity to recognize familiar patterns and movements of the brightest heavenly bodies.

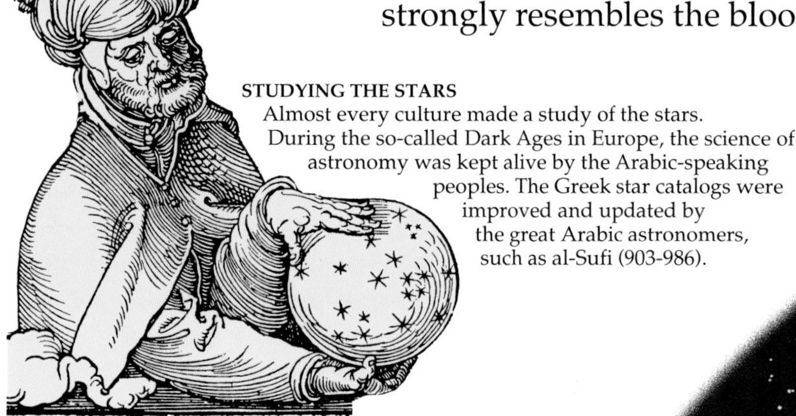

STUDYING THE STARS
Almost every culture made a study of the stars. During the so-called Dark Ages in Europe, the science of astronomy was kept alive by the Arabic-speaking peoples. The Greek star catalogs were improved and updated by the great Arabic astronomers, such as al-Sufi (903-986).

An engraving of al-Sufi with a celestial globe

UNCHANGING SKY
In all but the largest cities, where the stars are shrouded by pollution or hidden by the glare of street lights, the recurring display of the night sky is still captivating. The view of the stars from the Earth has changed remarkably little during the past 10,000 years. The sky on any night in the 20th century is nearly the same as the one seen by people who lived thousands of years ago. The night sky for people of the early civilizations would have been more accessible because their lives were not as sheltered from the effects of nature as ours are. Despite the advances in the technology of astronomical observation, which include radio telescopes where the images appear on a computer screen, and telescopes launched into space to detect radiations that do not penetrate our atmosphere, there are still things the amateur astronomer can enjoy. Books and newspapers print star charts so that on a given night, in a specified geographical location, anyone looking upward into a clear sky can see the constellations for themselves.

FROM SUPERSTITION TO SCIENCE
The science of astronomy grew out of a belief in astrology (pp. 16-17), the power of the planets and stars to affect life on the Earth. Each planet was believed to have the personality and powers of one of the gods. Mars, the god of war, shown here, determined war, plague, famine, and violent death.

TRADITIONAL SYMBOLS
The heritage of the Greek science of the stars passed through many different civilizations. In each case, the figures of the constellations took on the personalities of the heroes of local legends. The Mediterranean animals of the zodiac were transformed by other cultures, such as the Persians and Indians, into more familiar creatures, like the ibex, Brahma bulls, or a crayfish. This page is from an 18th-century Arabic manuscript. It depicts the zodiacal signs of Gemini, Cancer, Aries, and Taurus. The signs are in the Arabic script, which is read from right to left.

Quetzalcoatl

AZTEC MYTHOLOGY
In the Americas, the mythology of the stars was stronger than it was in Europe and Asia. This Aztec calendar shows the god Quetzalcoatl, who combined the influences of the Sun and Venus. His worship included ritual human sacrifice.

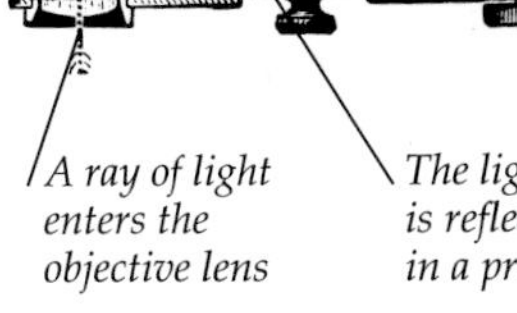

LOOKING AT STARS
Many of the sky's mysteries can be seen with a good pair of binoculars. This 20th-century pair gives a better view of the heavens than either Newton, Galileo, or other great astronomers could see with their best telescopes (pp. 20-21).

IMAGING SPACE
We can see 1,000 times farther into space than the ancients could. In 1990 NASA launched the first space telescope. The Hubble Space Telescope, orbiting beyond the Earth's atmosphere, can now produce high-resolution images of objects several billions of light-years (p. 60) away. Most of its time is spent pinpointing the areas near black holes and measuring the effect of gravity on starlight. Its faulty mirror and the solar panels were repaired in 1994 in a successful spacewalk from the Space Shuttle (p. 35).

Ancient astronomy

BY WATCHING THE CYCLIC MOTION of the Sun, the Moon, and the stars, early observers soon realized that these repeating motions could be used to fashion the sky into a clock (to tell the passage of the hours of the day or night) and a calendar (to mark the progression of the seasons). Ancient monuments, such as Stonehenge in the UK and the pyramids of the Maya in Central America, offer evidence that the basic components of observational astronomy have been known for at least 6,000 years. With few exceptions, all civilizations have believed that the steady movements of the sky are the signal of some greater plan. The phenomenon of a solar eclipse (pp. 38-39), for example, was believed by some ancient civilizations to be a dragon eating the Sun. A great noise would successfully frighten the dragon away.

DEFYING THE HEAVENS
The ancient poets warn that you should never venture out to sea until the constellation of the Pleiades rises with the Sun in early May. If the Soviet leader Mikhail Gorbachev and U.S. President George Bush had remembered their Greek poets, they would have known better than to try to meet on a boat in the Mediterranean in December 1989. Their summit was almost canceled because of bad weather.

PHASES OF THE MOON
The changing face of the Moon has always deeply affected people. The New Moon was considered the best time to start an enterprise, and the Full Moon was often feared as a time when spirits were free to roam. The word "lunatic" comes from the Latin name for the Moon, *luna*, because it was believed that the rays of the Full Moon caused insanity.

THE WORLD'S OLDEST OBSERVATORY
The earliest observatory to have survived is the Chomsung dae Observatory in Kyongju, Korea. A simple beehive structure, with a central opening in the roof, it resembles a number of prehistoric structures found all over the world. Many modern observatories (pp. 26-27) still have a similar roof opening.

NAMING THE PLANETS
The spread of knowledge tends to follow the two routes of trade and war. As great empires expanded, they brought their gods, customs, and learning with them. The earliest civilizations believed that the stars and planets were ruled by the gods. The Babylonians, for example, named each planet after the god that had most in common with that planet's characteristics. The Greeks and the Romans adopted the Babylonian system, replacing the names with those of their own gods. All the planet names can be traced directly to the Babylonian planet-gods: Nergal has become Mars, and Marduk has become the god Jupiter.

The Roman god Jupiter

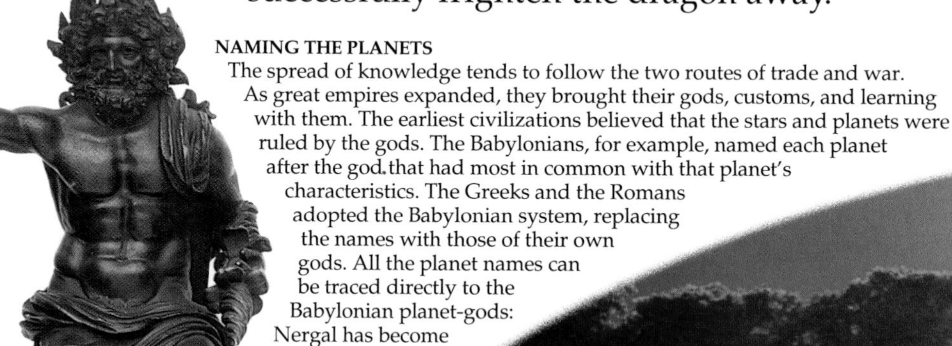

RECORDING THE SUN'S MOVEMENTS
Even though the precise significance of the standing stones at Stonehenge remains the subject of debate, it is clear from the arrangement of the stones that it was erected by prehistoric peoples specifically to record certain key celestial events, such as the summer and winter solstices and the spring and autumnal equinoxes. Although Stonehenge is the best known of the ancient megalithic monuments (those made of stone in prehistoric times), the sheer number of similar sites throughout the world underlines how many prehistoric peoples placed an enormous importance on recording the motions of the Sun and Moon.

BABYLONIAN RECORDS

The earliest astronomical records are in the form of clay tablets from ancient Mesopotamia and the great civilizations that flourished in the plains between the Tigris and Euphrates Rivers for more than 2,000 years. The oldest surviving astronomical calculations are relatively late, dating from the 4th century BC, but they are clearly based on generations of astronomical observations.

Back of a Persian astrolabe, 1707

THE ASTROLABE

One of the problems faced by ancient astronomers was how to simplify the complex calculations needed to predict the positions of the planets and stars. One useful tool was the astrolabe, whose different engraved plates reproduce the sphere of the heavens in two dimensions. The alidade with its sight holes is used to measure the height of the Sun or the stars. By setting this against the calendar scale on the outside of the instrument, a number of different calculations can be made.

PLANNING THE HARVEST

For nearly all ancient cultures the primary importance of astronomy was as a signal of seasonal changes. The Egyptians knew that when the star Sirius rose with the Sun, the annual flooding of the Nile was not far behind. Schedules for planting and harvesting were all set by the Sun, the Moon, and the stars.

Arabic manuscript from the 14th century showing an astrolabe being used

Ordering the Universe

A GREAT DEAL OF OUR KNOWLEDGE about the ancient science of astronomy comes from the Alexandrian Greek philosopher Claudius Ptolemaeus (c. AD 100-178), known as Ptolemy. He was an able scientist in his own right but, most importantly, he collected and clarified the work of all the great astronomers who had lived before him. He left two important sets of books. The *Almagest* was an astronomy textbook that provided an essential catalogue of all the known stars, updating Hipparchus. In the *Tetrabiblos*, Ptolemy discussed astrology. Both sets of books were the undisputed authority on their respective subjects for 1,600 years. Fortunately, they were translated into Arabic because, with the collapse of the Roman Empire around the 4th century, much accumulated knowledge disappeared as libraries were destroyed and books burned.

STAR CATALOGER
Hipparchus (190-120 BC) was one of the greatest of the Greek astronomers. He cataloged over 1,000 stars and developed the mathematical science of trigonometry. Here he is looking through a tube to help him concentrate on one area of the sky – the telescope was not yet invented (pp. 22-25).

Julius Caesar

THE LEAP YEAR
One of the problems confronting the astronomer-priests of antiquity was the fact that the lunar year and the solar year (p. 13) did not match up. By the middle of the 1st century BC, the Roman calendar was so mixed up that Julius Caesar (100-44 BC) ordered the Greek mathematician Sosigenes to develop a new system. He came up with the idea of a leap year every four years. This meant that the odd quarter day of the solar year was added in as one whole day once every four years.

Europe

Red Sea

Ocean

Africa

Facsimile (1908) of the Behaim terrestrial globe

FARNESE ATLAS
Very few images of the constellations have survived from antiquity. The main source for our knowledge is this 2nd-century Roman copy of an earlier Greek statue. The marble statue has the demi-god Atlas holding the heavens on his shoulders. All of the 48 Ptolemaic constellations are clearly marked in low relief.

SPHERICAL EARTH
The concept of a spherical Earth can be traced back to Greece in the 6th century BC. By Ptolemy's time, astronomers were accustomed to working with terrestrial and celestial globes. The first terrestrial globe to be produced since antiquity, the 15th-century globe by Martin Behaim, shows an image of the Earth that is half-based on myth. The Red Sea, for example, is colored red.

ARABIC SCHOOL OF ASTRONOMY
During the Dark Ages the great civilizations of Islam continued to develop the science of astronomy. Ulugh Beigh (c. 15th century) set up his observatory on this site in Samarkand, Central Asia, where measurements were made with the naked eye.

Geocentric Universe

It is logical to make assumptions from what your senses tell you. From the Earth it looks as if the heavens are circling over our heads. There is no reason to assume the Earth is moving at all. Ancient philosophers, naturally, believed that their Earth was stable and the center of the great cosmos. The planets were fixed to a series of crystalline spheres nested inside each other. The outermost sphere contained the stars.

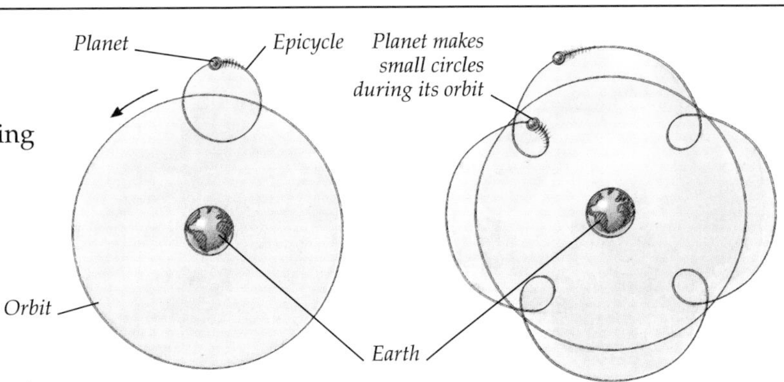

PROBLEMS WITH THE GEOCENTRIC UNIVERSE
The main problem with the model of an Earth-centered Universe was that it did not help explain the apparently irrational behavior of some of the planets, which sometimes appear to stand still or move backward against the background of the stars (p. 19). Early civilizations assumed that these odd movements were signals from the gods, but the Greek philosophers spent centuries trying to develop rational explanations for what they saw. The most popular was the notion of epicycles. The planets moved in small circles (epicycles) on their orbits as they circled the Earth.

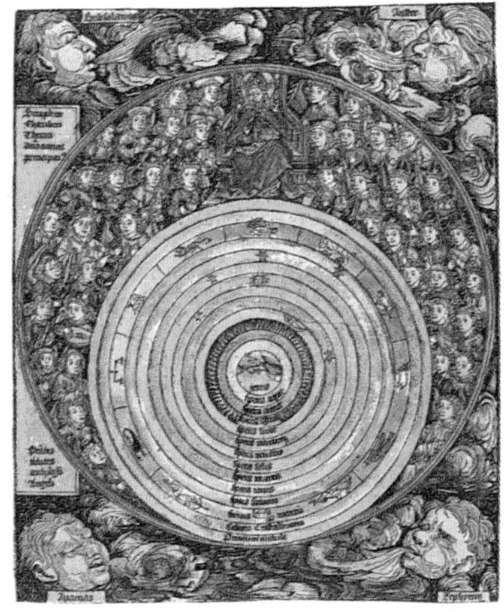

Engraving (1490) of the Ptolemaic Universe

THE EARTH AT THE CENTER
The geocentric or Earth-centered Universe is often referred to as the Ptolemaic Universe by later scholars to indicate that this was how classical scientists, like the great Ptolemy, believed the Universe was structured. He saw the Earth as the center of the Universe with the Moon, the known planets, and the Sun moving around it. Aristarchus (c. 310-230 BC) had already suggested that the Earth travels around the Sun, but his theory was rejected because it did not fit in with the mathematical and philosophical beliefs of the time.

TEACHING TOOL
Astronomers have always found it difficult to explain the three-dimensional motions of the heavens. Ptolemy used something like this armillary sphere to do his complex astronomical calculations and to pass these ideas on to his students.

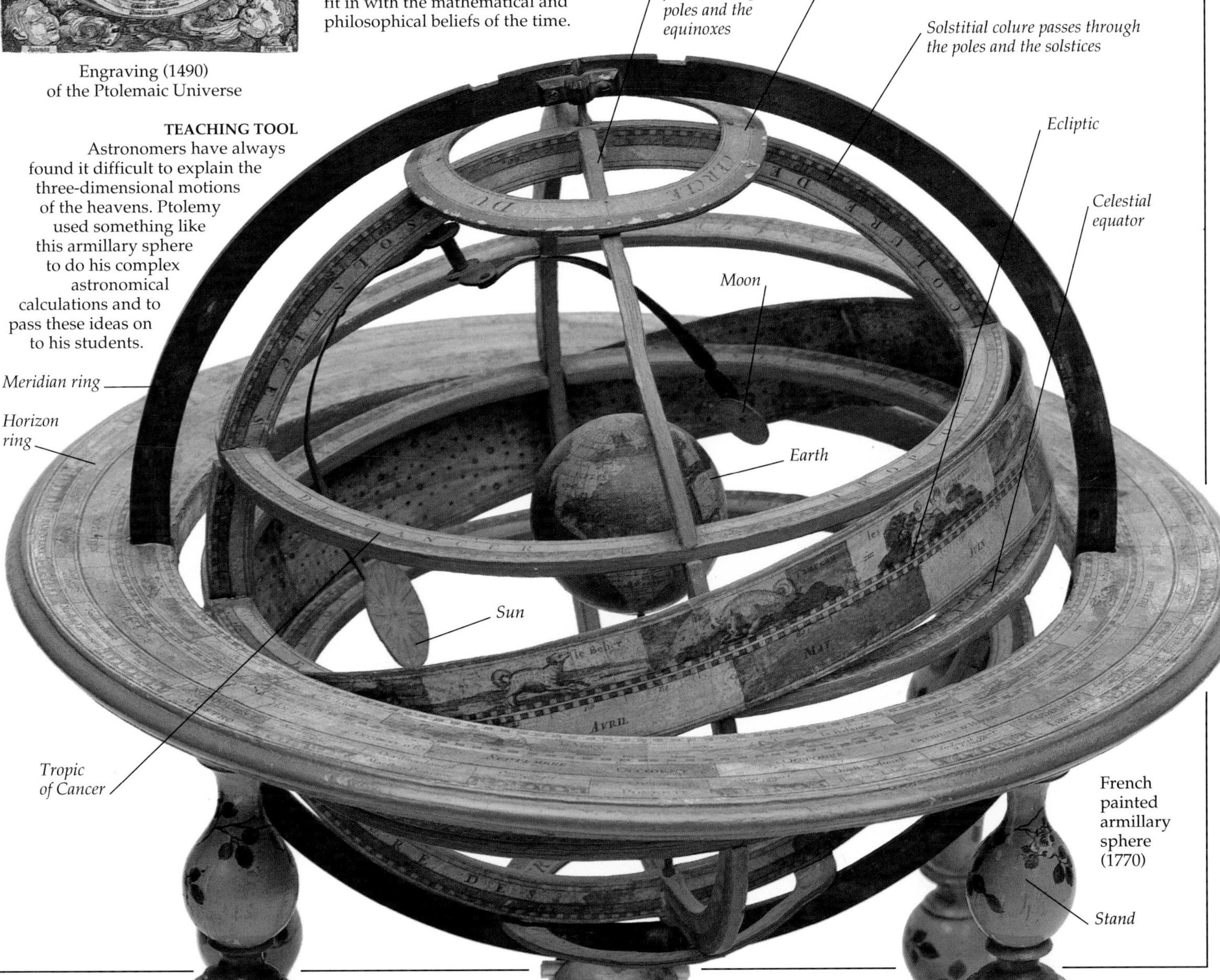

French painted armillary sphere (1770)

The celestial sphere

THE POSITIONS OF ALL OBJECTS IN SPACE are measured according to specific celestial coordinates. The best way to understand the cartography, or mapping, of the sky is to recall how the ancient philosophers imagined the Universe was shaped. They had no real evidence that the Earth moves, so they concluded that it was stationary and that the stars and planets revolve around it. They could see the stars wheeling around a single point in the sky and assumed that this must be one end of the axis of a great celestial sphere. They called it a crystalline sphere, or the sphere of fixed stars, because none of the stars seemed to change their positions relative to each other. The celestial coordinates used today come from this old-fashioned concept of a celestial sphere. The starry (celestial) and earthly (terrestrial) spheres share the same coordinates, such as a North and South Pole and an equator.

STAR TRAILS
A long photographic exposure of the sky taken from the northern hemisphere of the Earth shows the way in which stars appear to go in circles around the Pole Star or Polaris. Polaris is a bright star that lies within 1° of the true celestial pole, which, in turn, is located directly above the North Pole of the Earth. The rotation of the Earth on its north/south axis is the reason why the stars appear to move across the sky. Those closer to the Poles appear to move less than those further away.

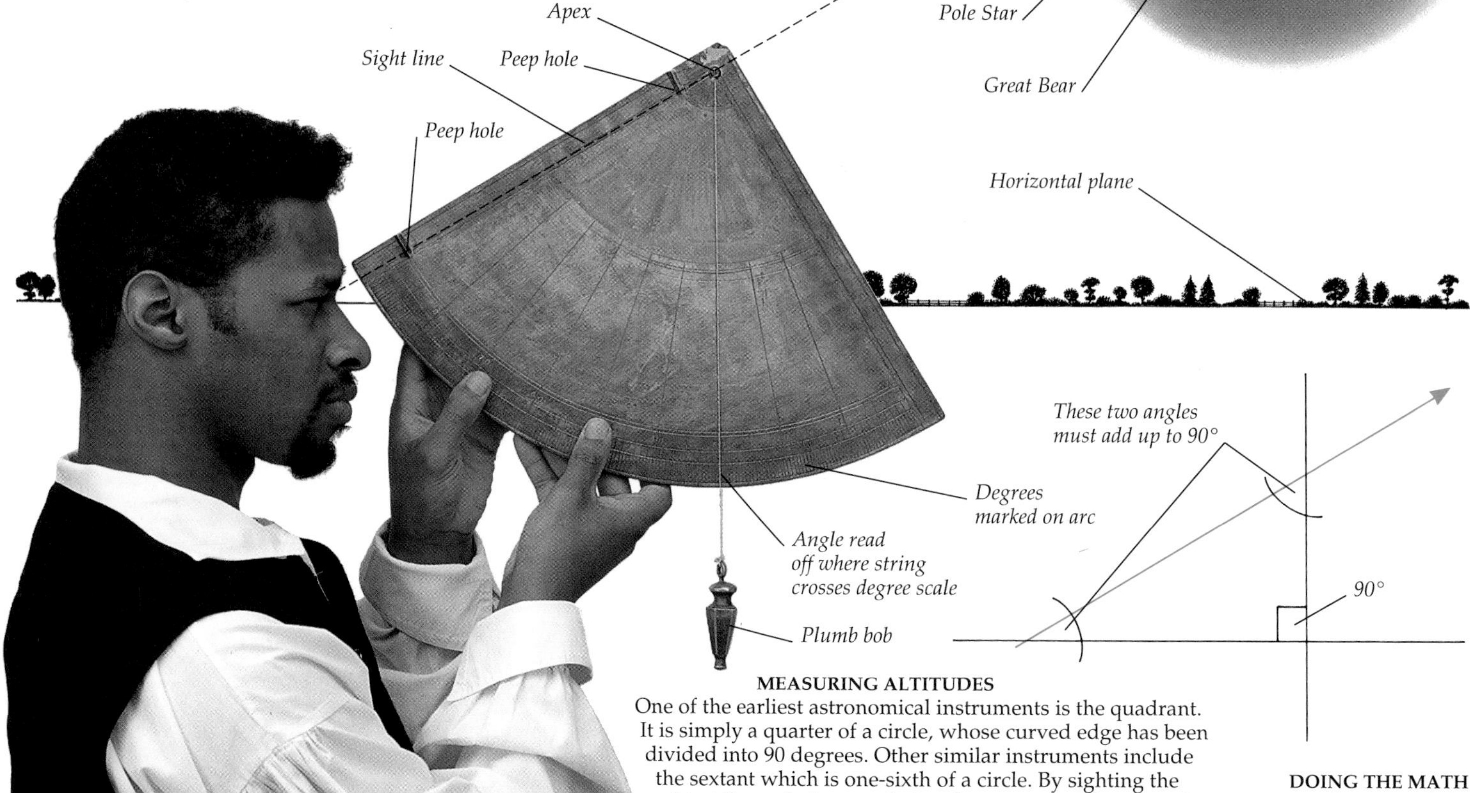

MEASURING ALTITUDES
One of the earliest astronomical instruments is the quadrant. It is simply a quarter of a circle, whose curved edge has been divided into 90 degrees. Other similar instruments include the sextant which is one-sixth of a circle. By sighting the object through the peep holes along one straight edge of the quadrant, the observer can measure the height, or altitude, of that object. The altitude is the height in degrees (°) of a star above the horizon; it is not a linear measurement. A string with a plumb bob falls from the apex of the quadrant so that it intersects the divided arc. Since the angle between the vertical of the plumb bob and the horizontal plane of the horizon is 90°, simple mathematics can be used to work out the angle of the altitude.

DOING THE MATH
The apex of the quadrant is a 90° angle. As the sum of the angles of a triangle adds up to 180°, this means that the sum of the other two angles must add up to 90° too.

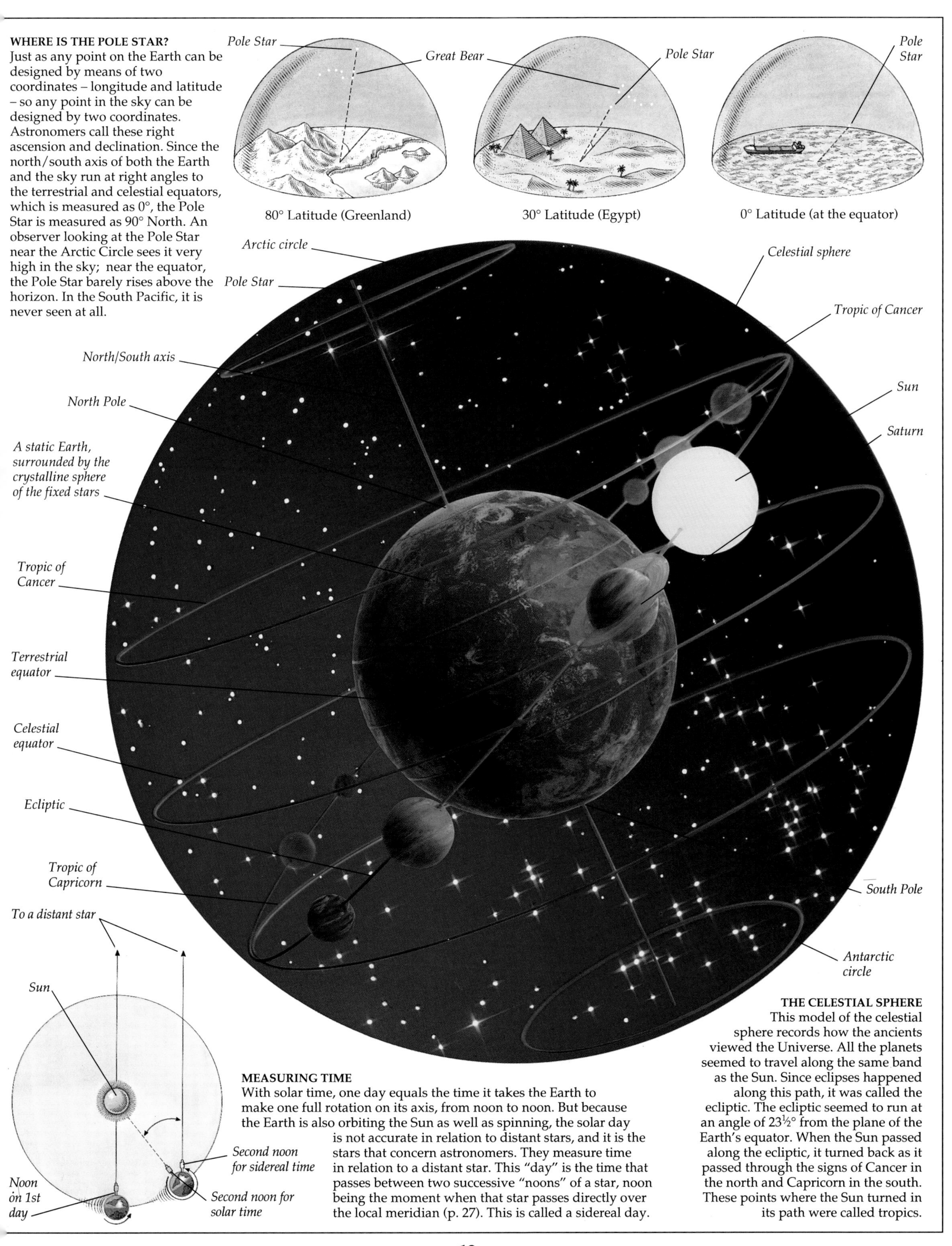

WHERE IS THE POLE STAR?
Just as any point on the Earth can be designed by means of two coordinates – longitude and latitude – so any point in the sky can be designed by two coordinates. Astronomers call these right ascension and declination. Since the north/south axis of both the Earth and the sky run at right angles to the terrestrial and celestial equators, which is measured as 0°, the Pole Star is measured as 90° North. An observer looking at the Pole Star near the Arctic Circle sees it very high in the sky; near the equator, the Pole Star barely rises above the horizon. In the South Pacific, it is never seen at all.

MEASURING TIME
With solar time, one day equals the time it takes the Earth to make one full rotation on its axis, from noon to noon. But because the Earth is also orbiting the Sun as well as spinning, the solar day is not accurate in relation to distant stars, and it is the stars that concern astronomers. They measure time in relation to a distant star. This "day" is the time that passes between two successive "noons" of a star, noon being the moment when that star passes directly over the local meridian (p. 27). This is called a sidereal day.

THE CELESTIAL SPHERE
This model of the celestial sphere records how the ancients viewed the Universe. All the planets seemed to travel along the same band as the Sun. Since eclipses happened along this path, it was called the ecliptic. The ecliptic seemed to run at an angle of 23½° from the plane of the Earth's equator. When the Sun passed along the ecliptic, it turned back as it passed through the signs of Cancer in the north and Capricorn in the south. These points where the Sun turned in its path were called tropics.

The uses of astronomy

WITH ALL THE TOOLS OF MODERN TECHNOLOGY, it is sometimes hard to imagine how people performed simple functions such as telling the time or knowing where they were on the Earth before the invention of the clock, maps, or navigational satellites. The only tools available were those provided by nature. The astronomical facts of the relatively regular interval of the day, the constancy of the movements of the fixed stars, and the assumption of certain theories, such as a spherical Earth, allowed people to measure their lives. By calculating the height of the Sun or certain stars, the ancient Greeks began to understand the shape and size of the Earth. In this way, they were able to determine their latitude (pp. 26-27). By plotting coordinates against a globe, they could fix their position on the Earth's surface. And by setting up carefully measured markers, or gnomons, they could begin to calculate the time of day.

Sun
Syene
Alexandria

MEASURING THE EARTH
About 230 BC, Eratosthenes (c. 270-190 BC) estimated the size of the Earth by using the Sun. He discovered that the Sun was directly above his head at Syene in Upper Egypt at noon on the summer solstice. In Alexandria, directly north, the Sun was about 7° from its highest point (the zenith) at the summer solstice. Since Eratosthenes knew that the Earth is spherical (360° in circumference), the distance between the two towns should be 7/360ths of the Earth's circumference.

Latitude scale
Sight hole
Movable cursor
Sight hole
Hour scale
Zodiac scale
Plumb bob

AN ANCIENT SUNDIAL
Very early on, people realized that they could keep time by the Sun. Simple sundials like this allowed the traveler or merchant to know the local time for several different towns during a journey. The altitude of the Sun was measured through the sight holes in the bow and stern of the "little ship." When the cursor on the ship's mast was set to the correct latitude, the plumb bob would fall on the proper time.

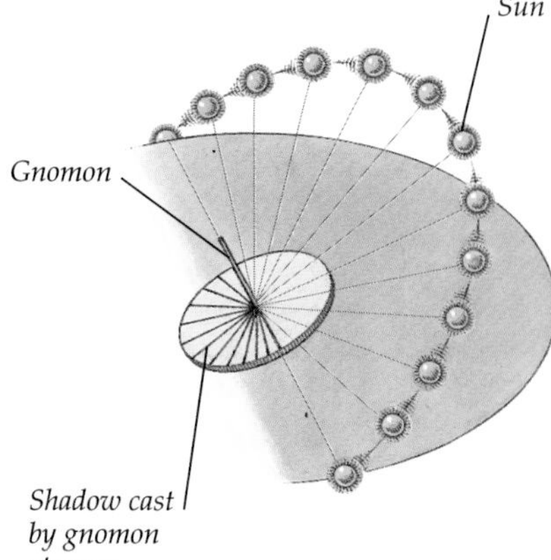

HOW A SUNDIAL WORKS
As the Sun travels across the sky, the shadow it casts changes in direction and length. A sundial works by setting a gnomon, or "indicator," so that the shadow the Sun casts at noon falls due north/south along a meridian. (A meridian is an imaginary line running from pole to pole; another name for meridian is a line of longitude.) The hours can then be divided before and after the noon mark. The terms AM and PM for morning and afternoon come from the Latin words meaning before and after the Sun passes the north/south meridian (*ante meridiem* and *post meridiem*).

Towns with their latitudes
Qiblah lid

FINDING MECCA
Islamic worship includes regular prayers in which the faithful face toward the holy city of Mecca. The qiblah indicator is a sophisticated instrument, developed during the Middle Ages to find the direction of Mecca. It also uses the Sun to determine the time for beginningand ending prayers.

TRAVELING TO THE SOUTH PACIFIC
It was thought that the early indigenous peoples of Polynesia were too "primitive" to have sailed the great distances between the north Pacific Ocean and New Zealand in the south. However, many tribes, including the Maoris of New Zealand, were capable of navigating thousands of miles using only the stars to guide them.

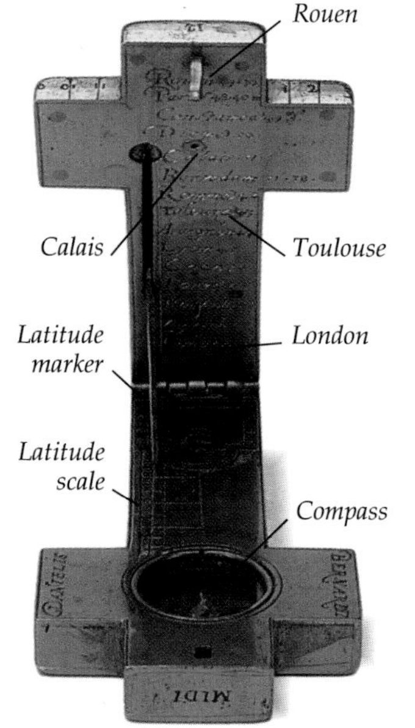

CRUCIFORM SUNDIAL
Traveling Christian pilgrims often worried that any ornament might be considered a symbol of vanity. They solved this problem by incorporating religious symbolism into their sundials. This dial, shaped in the form of a cross, provided the means for telling the time in a number of English and French towns.

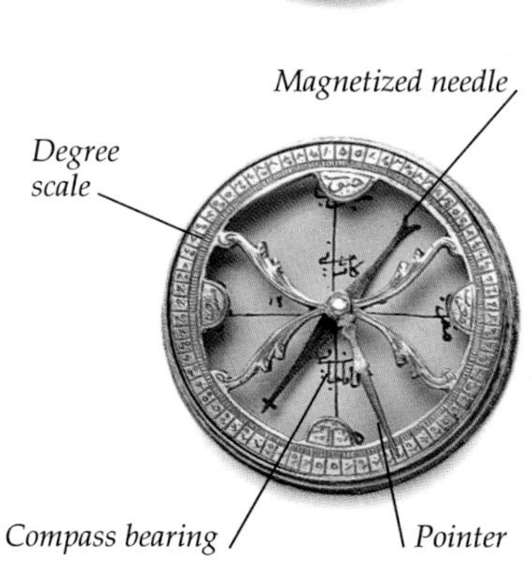

A CELESTIAL GLOBE

The celestial globe records the figures and stars of all the constellations against a grid of lines representing longitude and latitude. During the 17th and 18th centuries, all ships of the Dutch East India Company were given a matching pair of globes – terrestrial (p. 10) and celestial. Calculations could be made by comparing the coordinates on the two different globes. In practice, however, most navigators seemed to use flat sea charts to plot their journeys.

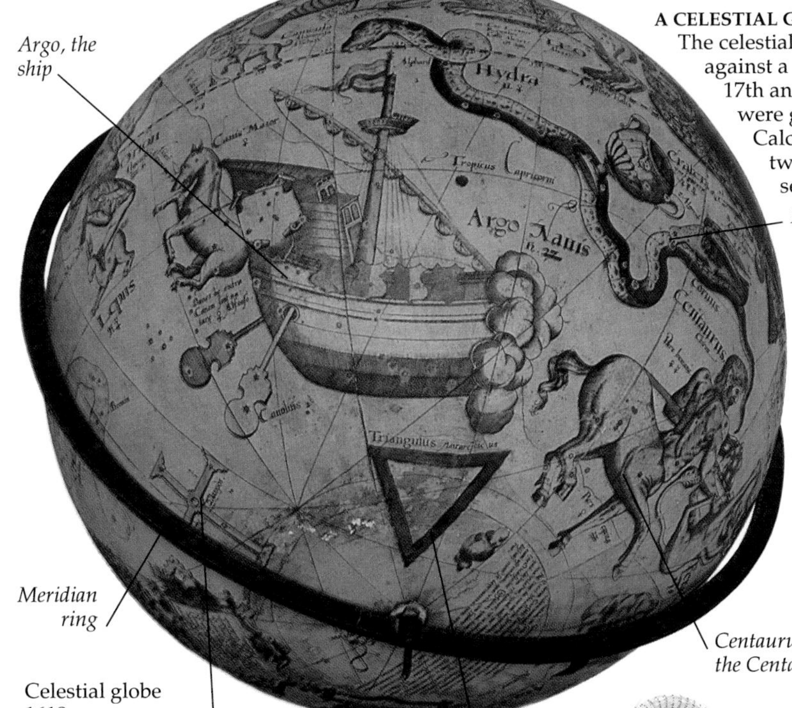

Argo, the ship

Hydra, the water snake

Meridian ring

Centaurus, the Centaur

The Southern Cross

Antarctic triangle

Celestial globe 1618

THE GREAT NAVIGATORS

Explorers of the 16th century had no idea what they would find when they set out to sea. Their heads were full of fables about mermaids and sea monsters. Even though this engraving of the Portuguese navigator Ferdinand Magellan (1480-1521) has many features that are clearly fantastical, it does show him using a pair of dividers to measure off an armillary sphere (p. 11). Beside the ship, the Sun God Apollo shines brightly; it was usually the Sun's position in the sky that helped a navigator find his latitude.

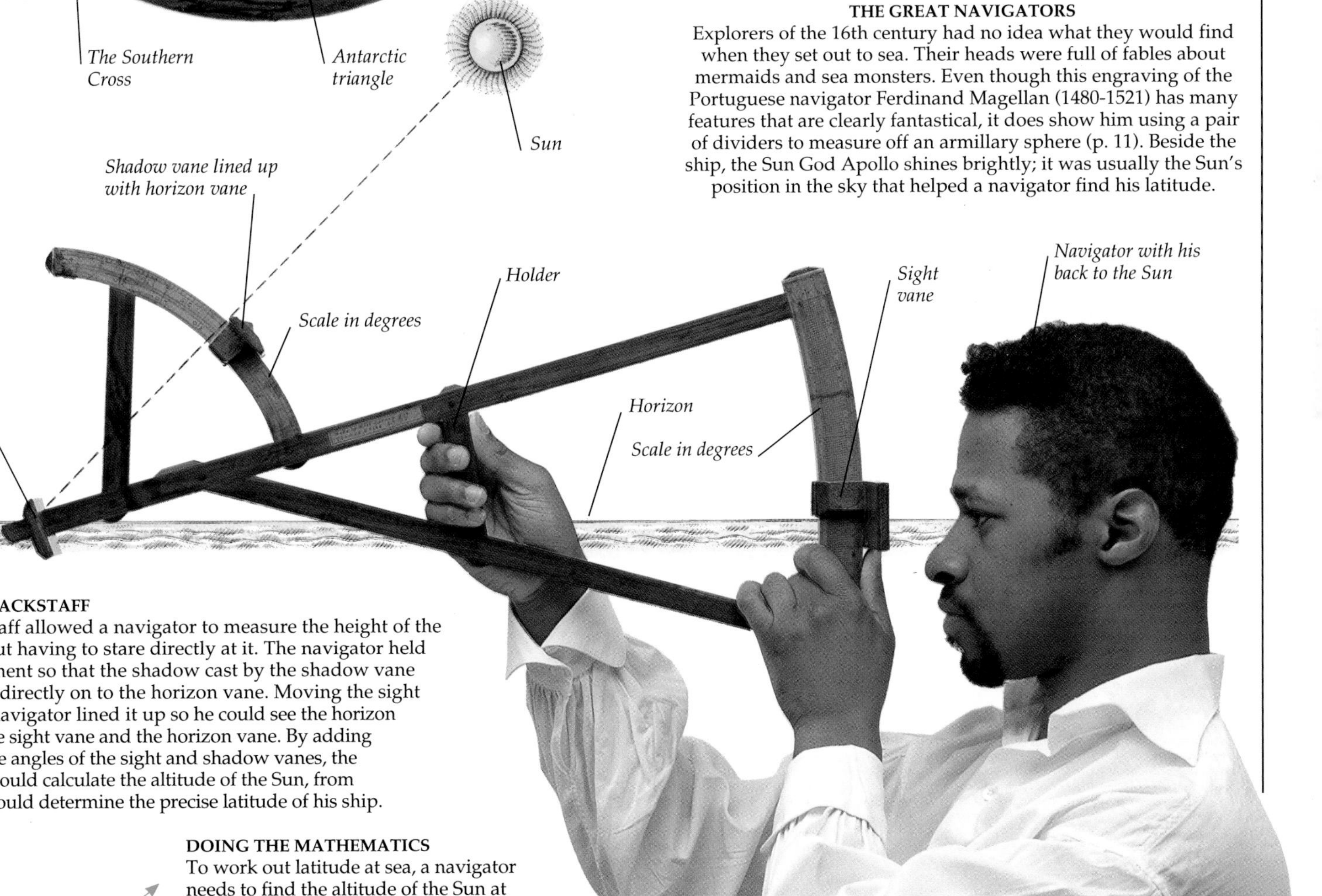

Sun

Shadow vane lined up with horizon vane

Horizon vane

Scale in degrees

Holder

Sight vane

Navigator with his back to the Sun

Horizon

Scale in degrees

USING A BACKSTAFF

The backstaff allowed a navigator to measure the height of the Sun without having to stare directly at it. The navigator held the instrument so that the shadow cast by the shadow vane would fall directly on to the horizon vane. Moving the sight vane, the navigator lined it up so he could see the horizon through the sight vane and the horizon vane. By adding together the angles of the sight and shadow vanes, the navigator could calculate the altitude of the Sun, from which he could determine the precise latitude of his ship.

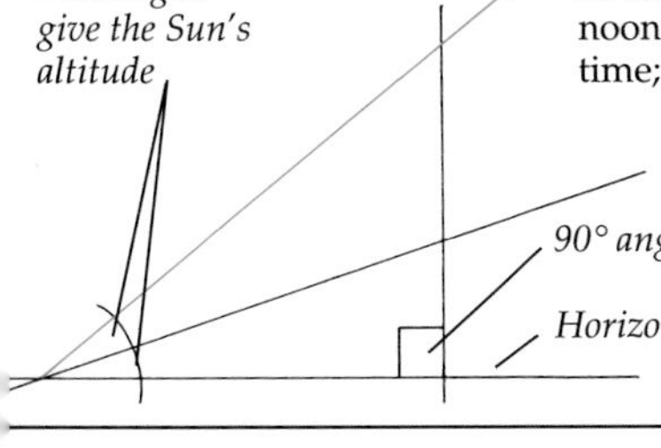

DOING THE MATHEMATICS

To work out latitude at sea, a navigator needs to find the altitude of the Sun at noon. He doesn't even need to know the time; as long as the Sun is at its highest point in the sky, the altitude can be measured with a backstaff or other instrument (p. 12). Then, using nautical tables of celestial coordinates, he can find his latitude with a simple equation using the angle of altitude and the coordinates of the Sun in the celestial sphere (p. 13).

Two angles give the Sun's altitude

90° angle

Horizon

Astrology

THE WORD "ASTROLOGY" comes from the Greek *astron* meaning "star" and *logos* meaning "the science." Since Babylonian times, people staring at the night sky were convinced that the regular motions of the heavens were indications of some great cosmic purpose. Priests and philosophers believed that if they could map the stars and their movements, they could decode these messages and understand the patterns that had an effect on past and future events. What was originally observational astronomy – observing the stars and planets – gradually grew into the astrology that has today become a regular part of many people's lives. However, there is no evidence that the stars and planets have any effect on our personalities or our destinies. Astronomers now think astrology is superstition. Its original noble motives should not be forgotten, however. For most of the so-called Dark Ages, when all pure science was in deep hibernation, it was astrology and the desire to know about the future that kept the science of astronomy alive.

THE ASTROLOGER
In antiquity, the astrologer's main task was to predict the future. This woodcut, dating from 1490, shows two astrologers working with arrangements of the Sun, Moon, and planets to find the astrological effects on peoples' lives.

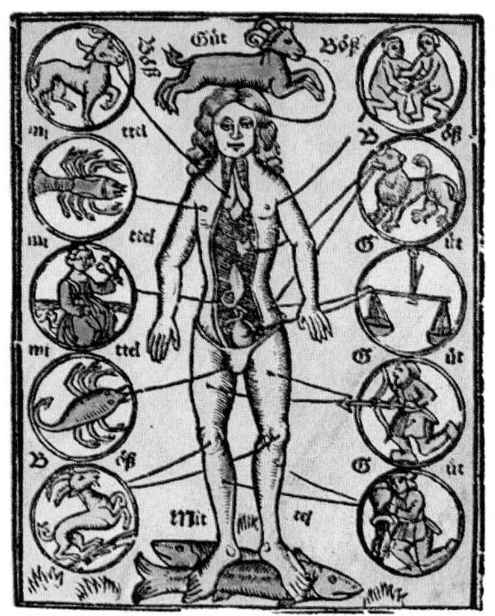

RULERSHIP OVER ORGANS
Until the discoveries of modern medicine, people believed that the body was governed by four different types of essences called "humors." An imbalance in these humors would lead to illness. Each of the 12 signs of the zodiac (right) had special links with each of the humors and with parts of the human body. So, for example, for a headache due to moisture in the head (a cold), treatment would be with a drying agent – some plant ruled by the Sun or an "Earth sign," like Virgo – when a New Moon was well placed toward the sign of Aries, which ruled the head.

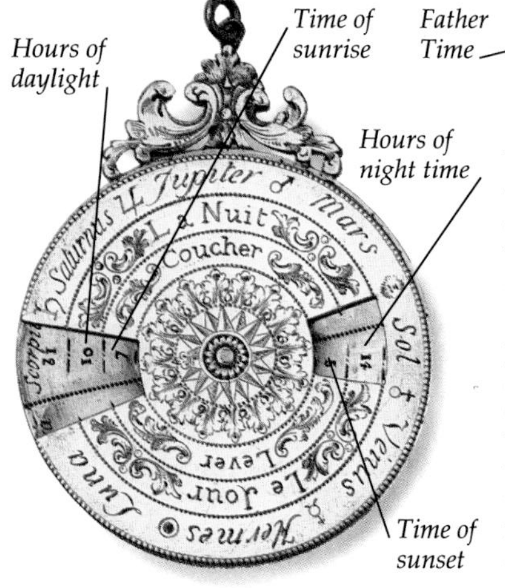

PERPETUAL CALENDAR
The names for the days of the week show traces of astrological belief – for example, Sunday is the Sun's day, and Monday is the Moon's day. This simple perpetual calendar, which has small planetary signs next to each day, shows the day of the week for any given date. The user can find the day by turning the inner dial to a given month or date and reading off the information.

LEO, THE LION
These 19th-century French constellation cards show each individual star marked with a hole through which light shines. Astrologically, each zodiacal sign has its own properties and its own friendships and enemies within the zodiacal circle. Each sign is also ruled by a planet, which similarly has its own properties, friendships, and enemies. So, for example, a person born while the Sun is passing through Leo is supposed to be kingly, like a lion.

PLANETARY POSITIONS

One way in which planets are supposed to be in or out of harmony with one another depends on their relative positions in the heavens. When two planets are found within a few degrees of each other, they are said to be in conjunction. When planets are separated by exactly 180° in the zodiacal band, they are said to be in opposition.

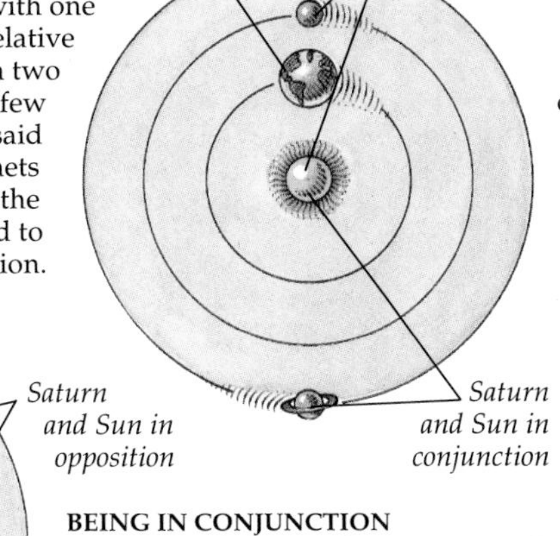

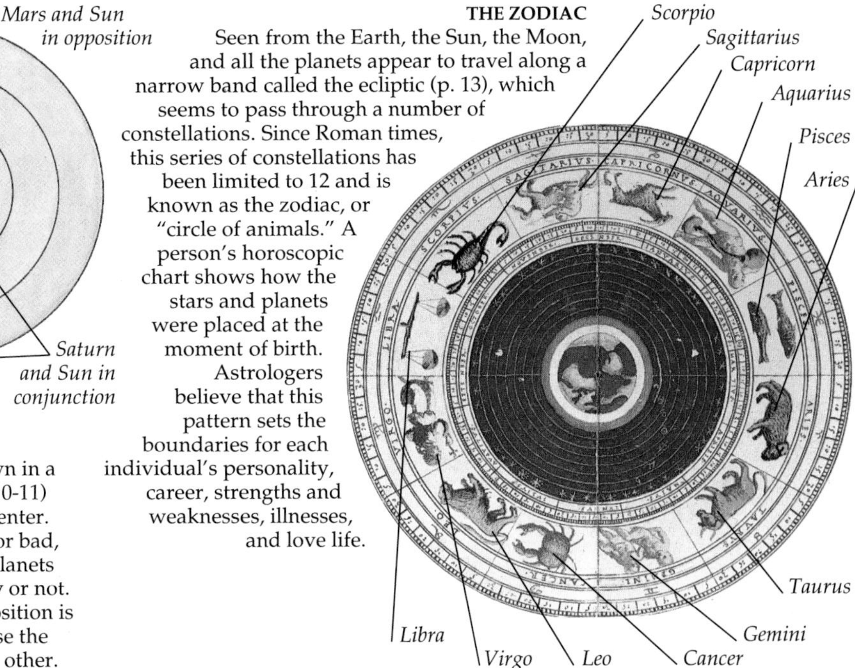

THE ZODIAC

Seen from the Earth, the Sun, the Moon, and all the planets appear to travel along a narrow band called the ecliptic (p. 13), which seems to pass through a number of constellations. Since Roman times, this series of constellations has been limited to 12 and is known as the zodiac, or "circle of animals." A person's horoscopic chart shows how the stars and planets were placed at the moment of birth. Astrologers believe that this pattern sets the boundaries for each individual's personality, career, strengths and weaknesses, illnesses, and love life.

BEING IN CONJUNCTION

The planets here are shown in a geocentric Universe (pp. 10-11) where the Earth is at the center. Conjunctions can be good or bad, depending on whether the planets involved are mutually friendly or not. Astrologers believe that an opposition is malefic, or "evil-willing," because the planets are fighting against each other.

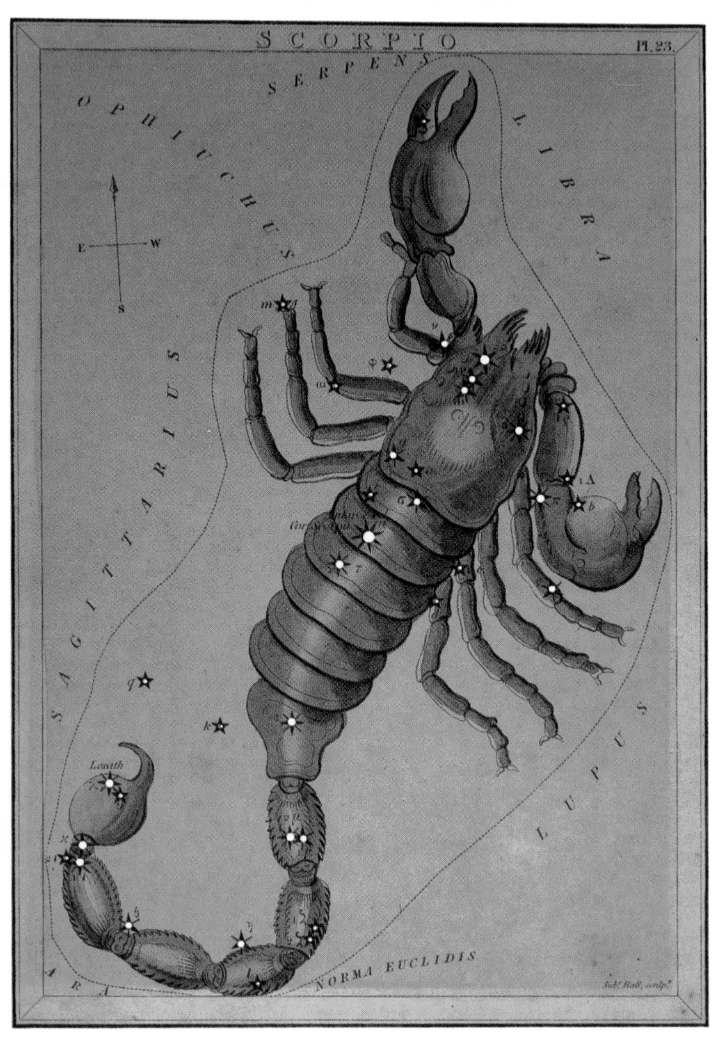

SCORPIO, THE SCORPION

Most of the constellations are now known by the Latinized versions of their original Greek names. This card shows Scorpius, or Scorpio. This is the sign "through which" the Sun passes between late October and late November. Astrologers believe that people born during this time of year are intuitive, yet secretive, like a scorpion scuttling under a rock.

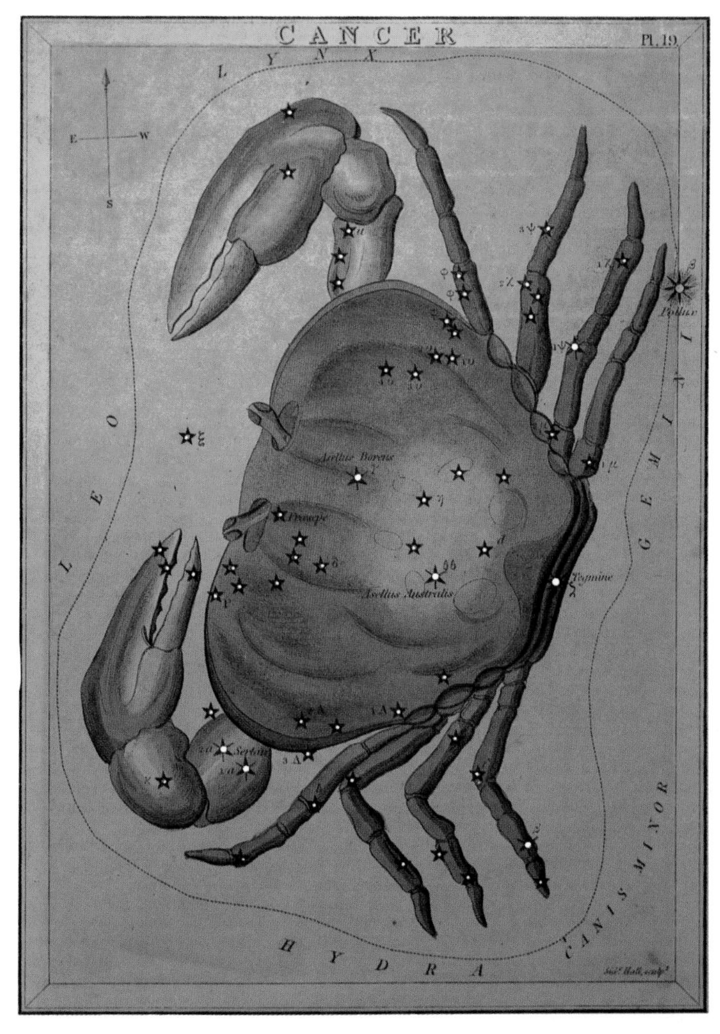

CANCER, THE CRAB

Someone who is born while the Sun is transitting the constellation Cancer is supposed to be a homebody, like a crab in its shell. These hand-painted cards are collectively known as *Urania's Mirror* – Urania is the name of the muse of astronomy (p. 19). By holding the cards up to the light, it is possible to learn the shapes and relative brightness of the stars in each constellation.

The Copernican revolution

In 1543 Nicolaus Copernicus published a book that changed the perception of the Universe. In his *De revolutionibus orbium coelestium* ("Concerning the revolutions of the celestial orbs"), Copernicus argued that the Sun, and not the Earth, is at the center of the Universe. It was a heliocentric Universe, *helios* being the Greek word for Sun. His reasoning was based on the logic of the time. He argued that a sphere moves in a circle that has no beginning and no end. Since the Universe and all the heavenly bodies are spherical, their motions must be circular and uniform. In the Ptolemaic, Earth-centered system (pp. 10-11), the paths of the planets are irregular. Copernicus assumed that uniform motions in the orbits of the planets appear irregular to us because the Earth is not at the center of the Universe. These discoveries were put forward by many different astronomers, but they ran against the teachings of both the Protestant and Catholic churches. In 1616 all books written by Copernicus and any others that put the Sun at the center of the Universe were condemned by the Catholic Church.

NICOLAUS COPERNICUS
The Polish astronomer Nicolaus Copernicus (1473-1543) made few observations. Instead, he read the ancient philosophers and discovered that none of them had been able to agree about the structure of the Universe.

COPERNICAN UNIVERSE
Copernicus based the order of his Solar System on how long it took each planet to complete a full orbit. This early print shows the Earth in orbit around the Sun with the Zodiac beyond.

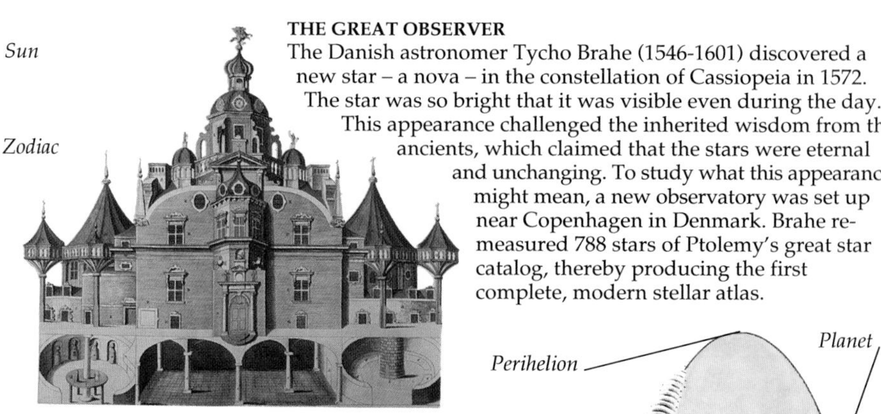

THE GREAT OBSERVER
The Danish astronomer Tycho Brahe (1546-1601) discovered a new star – a nova – in the constellation of Cassiopeia in 1572. The star was so bright that it was visible even during the day. This appearance challenged the inherited wisdom from the ancients, which claimed that the stars were eternal and unchanging. To study what this appearance might mean, a new observatory was set up near Copenhagen in Denmark. Brahe re-measured 788 stars of Ptolemy's great star catalog, thereby producing the first complete, modern stellar atlas.

Uranibourg, Tycho's observatory on the island of Hven

DRAWING AN ELLIPSE
An ellipse can be drawn by pushing two pins into a board and linking them with a loop of thread. When a pencil is placed within the loop and moved around the pins, keeping the loop tight, the shape it makes is an ellipse. The position of each pin is called a focus. In the Solar System the Sun is at one focus of the ellipse in a planetary orbit. The wider apart the pins are placed, the more eccentric the planetary orbit (pp. 36-37).

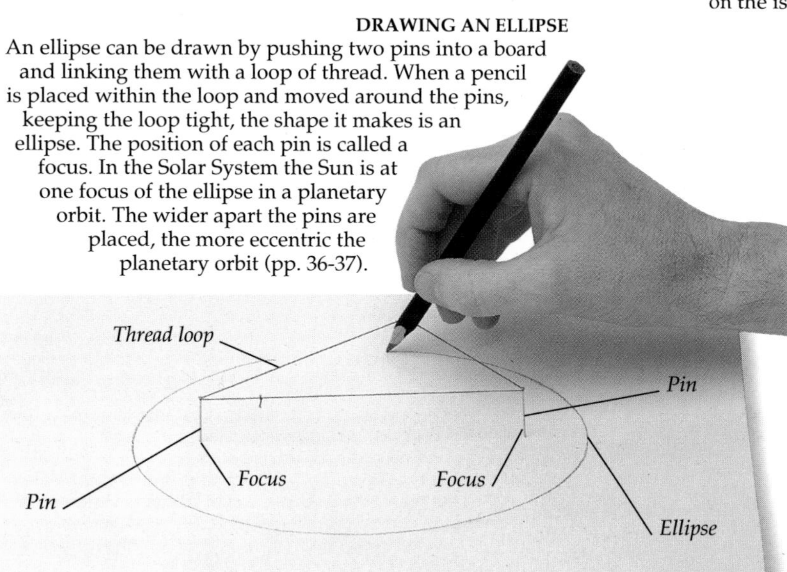

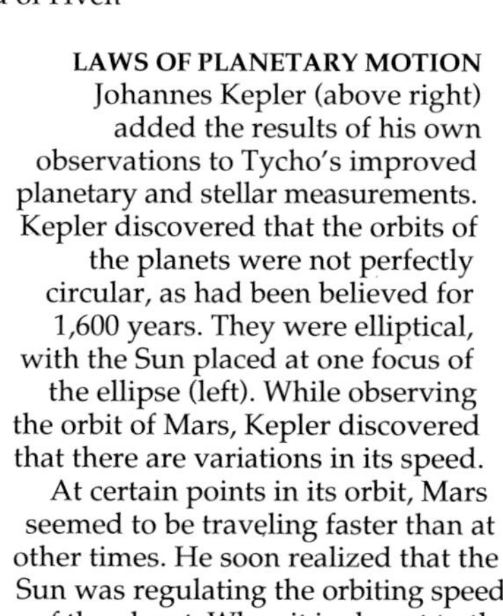

LAWS OF PLANETARY MOTION
Johannes Kepler (above right) added the results of his own observations to Tycho's improved planetary and stellar measurements. Kepler discovered that the orbits of the planets were not perfectly circular, as had been believed for 1,600 years. They were elliptical, with the Sun placed at one focus of the ellipse (left). While observing the orbit of Mars, Kepler discovered that there are variations in its speed. At certain points in its orbit, Mars seemed to be traveling faster than at other times. He soon realized that the Sun was regulating the orbiting speed of the planet. When it is closest to the Sun – its perihelion – the planet orbits most quickly; at its aphelion – farthest from the Sun – it slows down.

JOHANNES KEPLER (1571-1630)
It was due to the intervention of Tycho Brahe that the German mathematician Johannes Kepler landed the prestigious position of Imperial Mathematician in 1601. Tycho left all his papers to Kepler, who was a vigorous supporter of the Copernican heliocentric system. Kepler formulated three laws of planetary motion and urged Galileo (pp. 20) to publish his research in order to help prove the Copernican thesis.

Projected images in a planetarium

A model showing the true and apparent orbits of Mars from an earthly perspective

APPARENT PATHS
The irregular motion that disproved the geocentric universe was the retrograde motion of the planets. From an earthly perspective, some of the planets – particularly Mars – seem to double back on their orbits, making great loops in the night sky. (The light display above draws the apparent orbit of Mars.) Ptolemy proposed that retrograde motion could be explained by planets travelling on smaller orbits (p. 11). Once astronomers realized that the Sun is the center of the Solar System, the apparent path of Mars, for example, could be explained. But first it had to be understood that the Earth has a greater orbiting speed than that of Mars, which appeared to slip behind. Even though the orbit of Mars seems to keep pace with the Earth (below left), the apparent path is very different (above left).

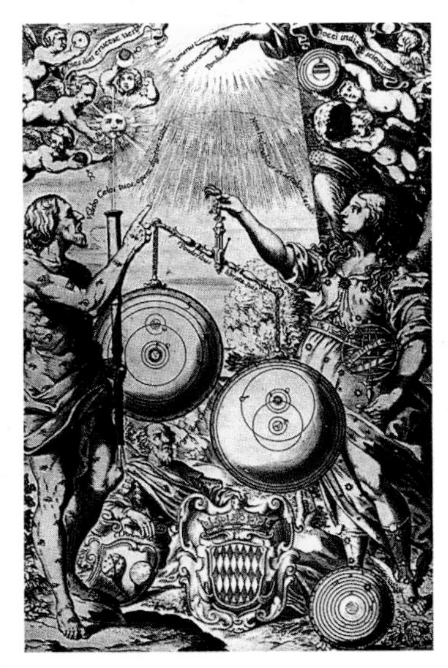

WEIGHING UP THE THEORIES
This engraving from a 17th-century manuscript shows Urania, the muse of astronomy, comparing the different theoretical systems for the arrangement of the Universe. Ptolemy's system is at her feet, and Kepler's is outweighed by Tycho's system on the right.

Intellectual giants

IT TAKES BOTH LUCK AND COURAGE to be a radical thinker. Galileo Galilei (1564-1642) had the misfortune of being brilliant at a time when new ideas were considered dangerous. His numerous discoveries, made with the help of the newly invented telescope, provided ample support for the Copernican heliocentric, or Sun-centered, Universe (pp. 18-19). Galileo's findings about the satellites of Jupiter (p. 50) and the phases of Venus clearly showed that the Earth could not be the center of all movement in the Universe and that the heavenly bodies were not perfect in their behavior. For this Galileo was branded a heretic and sentenced to a form of life imprisonment. The great English physicist Isaac Newton (1642-1727), born the year Galileo died, had both luck and courage. He lived in an age enthusiastic for new ideas, especially those related to scientific discovery.

GALILEO'S TELESCOPE
Galileo never claimed to have invented the telescope. In *Il saggiatore*, "The Archer," he commends the "simple spectacle-maker" who "found the instrument" by chance. When he heard of Lippershey's results (p. 22), Galileo reinvented the instrument from the description of its effects. His first telescope magnified at eight times. Within a few days, however, he had constructed a telescope with x 20 magnification. He went on to increase his magnification to x 30, having ground the lenses himself.

POPE URBAN VIII
Originally, the Catholic Church had welcomed Copernicus's work (pp. 18-19). However, by 1563 the Church was becoming increasingly strict and abandoned its previously lax attitude toward any deviation from established doctrine. Pope Urban VIII was one of the many caught in this swing. As a cardinal, he had been friendly with Galileo and often had Galileo's book, *Il saggiatore*, read to him aloud at meals. In 1635, however, he authorized the Grand Inquisition of Galileo's research.

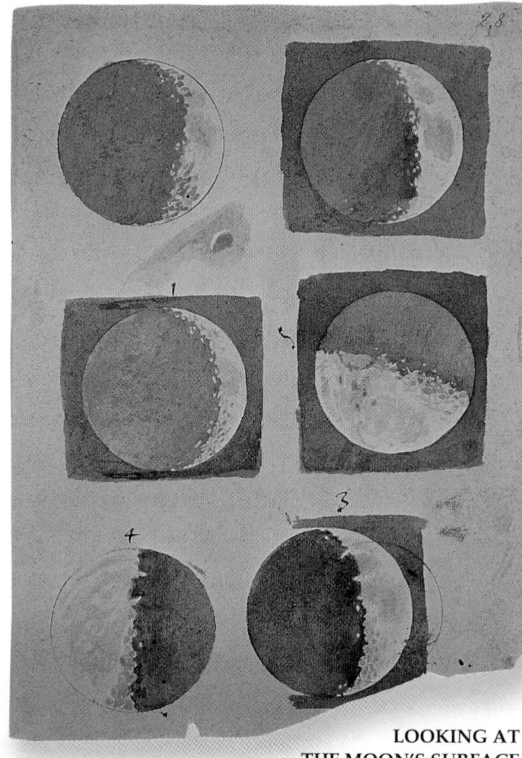

LOOKING AT THE MOON'S SURFACE
Through his telescope, Galileo measured the shadows on the Moon to show how the mountains there were much taller than those on Earth. These ink sketches were published in his book *Sidereus nuncius* "Messenger of the Stars" in 1610.

RENAISSANCE MAN
In 1611 Galileo traveled to Rome to discuss his findings about the Sun and its position in the Universe with the leaders of the Church. They accepted his discoveries, but not the theory that underpinned them – the Copernican, heliocentric Universe (pp. 18-19). Galileo was accused of heresy and, in 1635, condemned for disobedience and sentenced to house arrest until his death in 1642. He was pardoned in 1992.

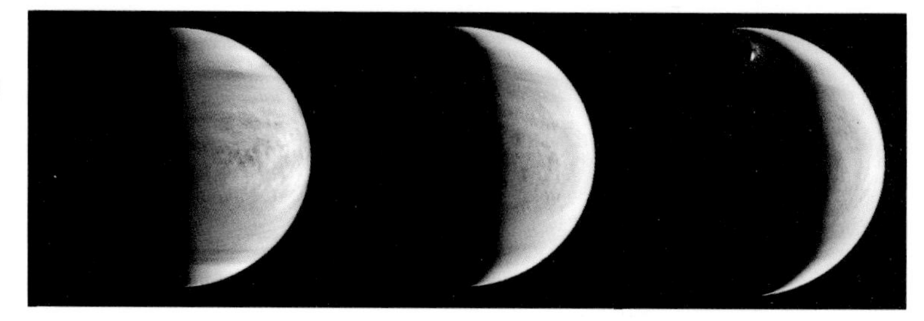

PHASES OF VENUS
From his childhood days, Galileo was characterized as the sort of person who was unwilling to accept facts without evidence. In 1610, by applying the telescope to astronomy, he discovered the moons of Jupiter and the phases of Venus. He immediately understood that the phases of Venus are caused by the Sun shining on a planet that revolves around it. He knew that this was proof that the Earth was not the center of the Universe. He hid his findings in a Latin anagram, or word puzzle, as he did with many of the discoveries that he knew would be considered "dangerous" by the authorities.

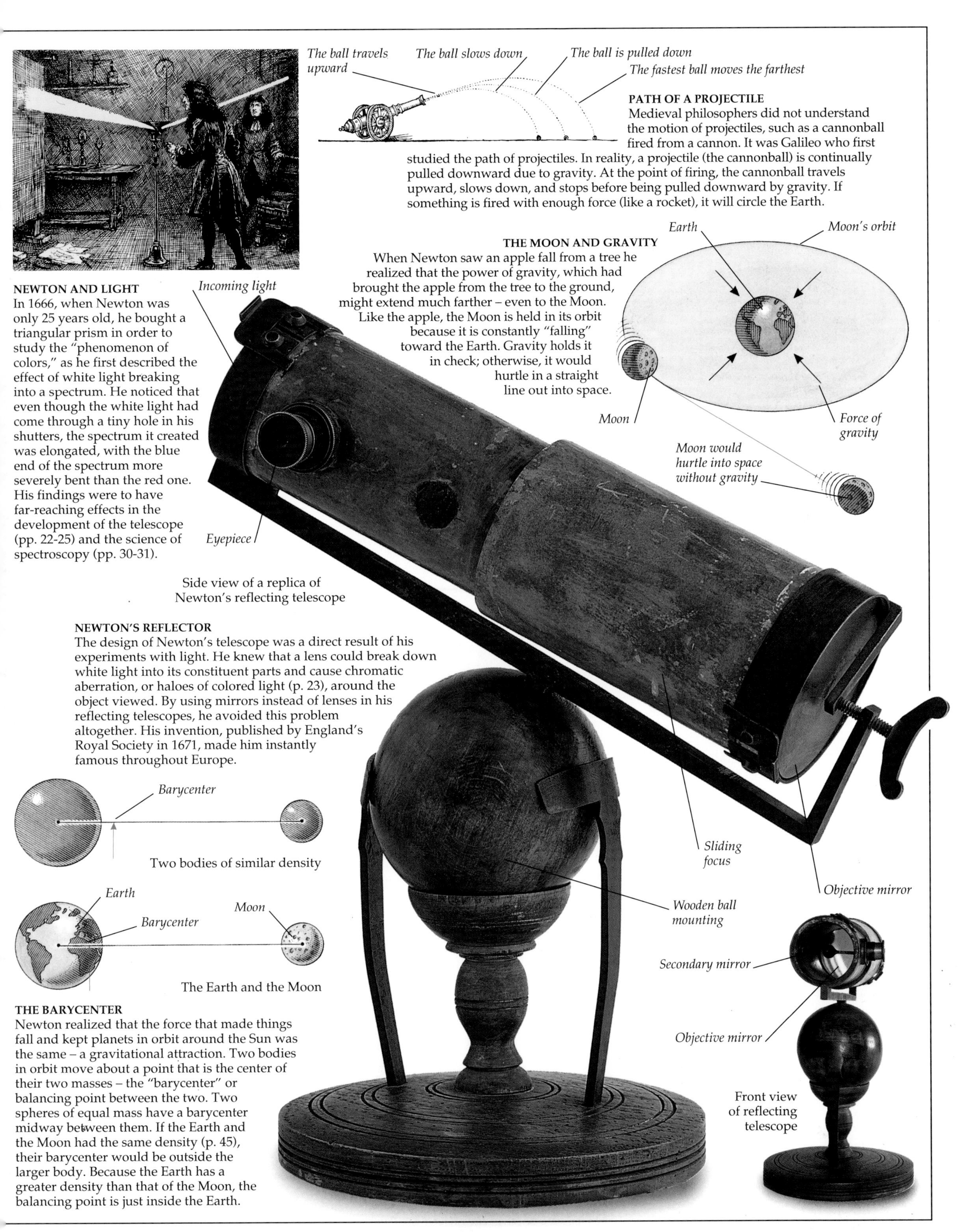

PATH OF A PROJECTILE
Medieval philosophers did not understand the motion of projectiles, such as a cannonball fired from a cannon. It was Galileo who first studied the path of projectiles. In reality, a projectile (the cannonball) is continually pulled downward due to gravity. At the point of firing, the cannonball travels upward, slows down, and stops before being pulled downward by gravity. If something is fired with enough force (like a rocket), it will circle the Earth.

NEWTON AND LIGHT
In 1666, when Newton was only 25 years old, he bought a triangular prism in order to study the "phenomenon of colors," as he first described the effect of white light breaking into a spectrum. He noticed that even though the white light had come through a tiny hole in his shutters, the spectrum it created was elongated, with the blue end of the spectrum more severely bent than the red one. His findings were to have far-reaching effects in the development of the telescope (pp. 22-25) and the science of spectroscopy (pp. 30-31).

THE MOON AND GRAVITY
When Newton saw an apple fall from a tree he realized that the power of gravity, which had brought the apple from the tree to the ground, might extend much farther – even to the Moon. Like the apple, the Moon is held in its orbit because it is constantly "falling" toward the Earth. Gravity holds it in check; otherwise, it would hurtle in a straight line out into space.

Side view of a replica of Newton's reflecting telescope

NEWTON'S REFLECTOR
The design of Newton's telescope was a direct result of his experiments with light. He knew that a lens could break down white light into its constituent parts and cause chromatic aberration, or haloes of colored light (p. 23), around the object viewed. By using mirrors instead of lenses in his reflecting telescopes, he avoided this problem altogether. His invention, published by England's Royal Society in 1671, made him instantly famous throughout Europe.

Two bodies of similar density

The Earth and the Moon

Front view of reflecting telescope

THE BARYCENTER
Newton realized that the force that made things fall and kept planets in orbit around the Sun was the same – a gravitational attraction. Two bodies in orbit move about a point that is the center of their two masses – the "barycenter" or balancing point between the two. Two spheres of equal mass have a barycenter midway between them. If the Earth and the Moon had the same density (p. 45), their barycenter would be outside the larger body. Because the Earth has a greater density than that of the Moon, the balancing point is just inside the Earth.

Optical principles

PEOPLE HAVE BEEN AWARE of the magnifying properties of a curved piece of glass since at least 2,000 BC. The Greek philosopher Aristophanes in the 5th century BC had used a glass globe filled with water in order to magnify the fine print in his manuscripts. In the middle of the 13th century the English scientist Roger Bacon (1214-1292) proposed that the "lesser segment of a sphere of glass or crystal" will make small objects appear clearer and larger. For this suggestion, Bacon's colleagues branded him a dangerous magician and imprisoned for 10 years. Even though eyeglasses were invented in Italy some time between 1285 and 1300, superstitions were not overcome for another 250 years when scientists discovered the combination of lenses that would lead to the invention of the telescope. There were two types of telescope. The refracting telescope uses lenses to bend light; the reflecting telescope uses mirrors to reflect the light back to the observer.

INVENTOR OF THE TELESCOPE
It is believed that the first real telescope was invented in 1608 in Holland by the glasses maker Hans Lippershey from Zeeland. According to the story, two of Lippershey's children were playing in his shop and noticed that by holding two lenses in a straight line they could magnify the weather vane on the local church. Lippershey placed the two lenses in a tube and claimed the invention of the telescope. In the mid 1550s an Englishman Leonard Digges had created a primitive instrument that, with a combination of mirrors and lenses, could reflect and enlarge objects viewed through it. There was controversy about whether this was a true scientific telescope or not. It was Galileo (p. 20) who adapted the telescope to astronomy.

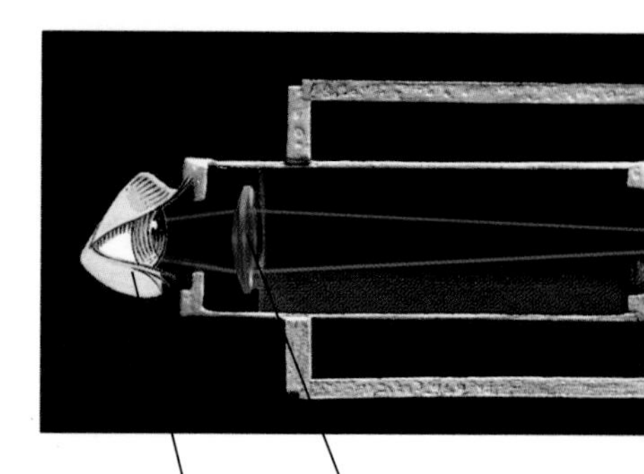

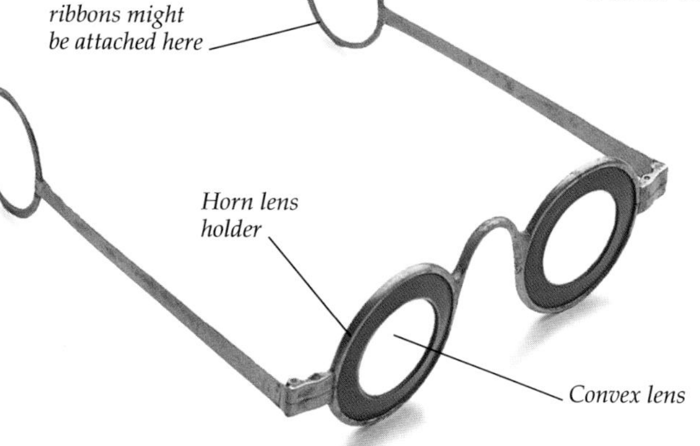

EARLY EYEGLASSES (1750)
Most early glasses like these had convex lenses. These helped people who were farsighted to focus on objects close at hand. Later, glasses were made with concave lenses for those who were nearsighted.

Water

Light from laser

Path of light is bent again on reentering air

Light is bent

HOW REFRACTION WORKS
Light usually travels in a straight line, but it can be bent or "refracted" by passing it through substances of differing densities. This laser beam (here viewed from overhead) seems to bend as it is directed at a rectangular-shaped container of water because the light is passing through three different media – water, glass, and air.

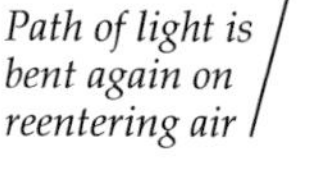

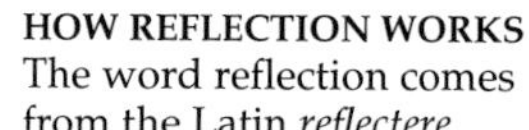

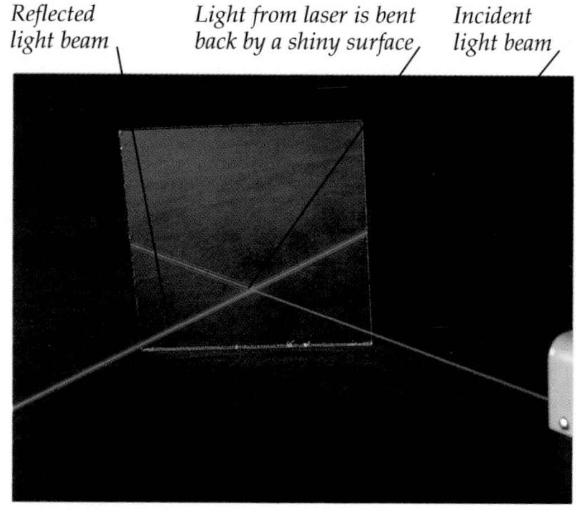

HOW REFLECTION WORKS
The word reflection comes from the Latin *reflectere*, meaning to "bend back." A shiny surface will bend back rays of light that strike it. The rays approaching the mirror are called incident rays and those leaving it are called outgoing, or reflected, rays. The angle at which the incident rays hit the mirror is the same as the angle of the reflected rays leaving it. What the eye sees are the light rays reflected in the mirror.

JOHN DOLLOND
The English optician John Dollond (1706-1761) was the first to perfect the achromatic lens so that it might be manufactured more easily and solve the problem of chromatic aberration. Dollond claimed to have invented a new method of refraction.

CHROMATIC ABERRATION
When light goes through an ordinary lens, each color in the spectrum is bent at a different angle causing rainbows to appear around the images viewed. The blue end of the spectrum will bend more sharply than the red end of the spectrum so that the two colors will focus at different points. This is chromatic aberration. By adding a second lens made from a different kind of glass (and with a different density), all the colors focus at the same point and the problem is corrected.

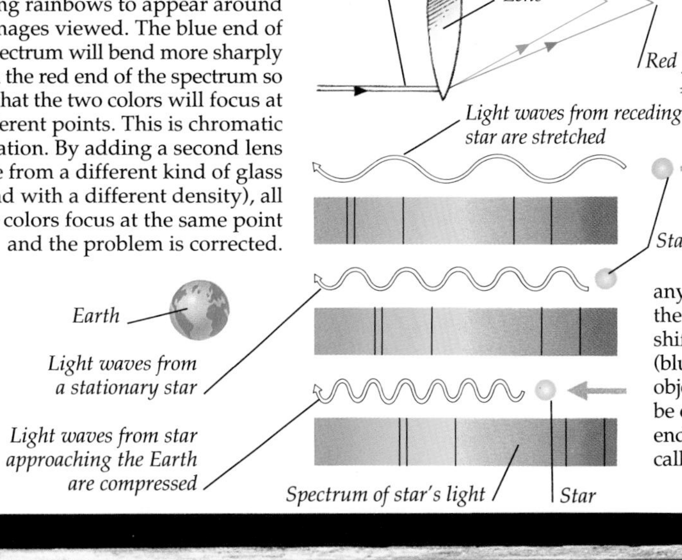

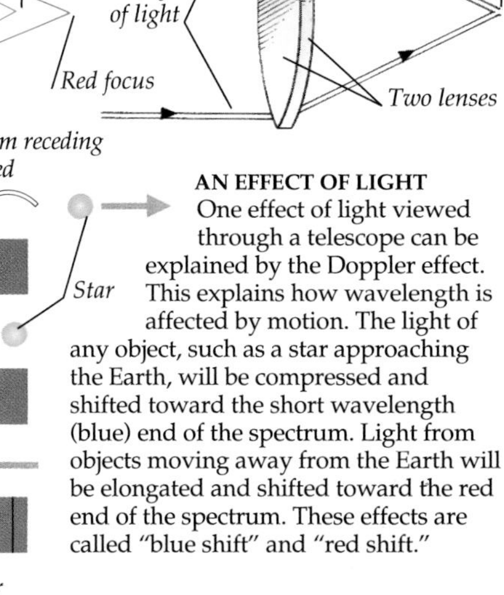

AN EFFECT OF LIGHT
One effect of light viewed through a telescope can be explained by the Doppler effect. This explains how wavelength is affected by motion. The light of any object, such as a star approaching the Earth, will be compressed and shifted toward the short wavelength (blue) end of the spectrum. Light from objects moving away from the Earth will be elongated and shifted toward the red end of the spectrum. These effects are called "blue shift" and "red shift."

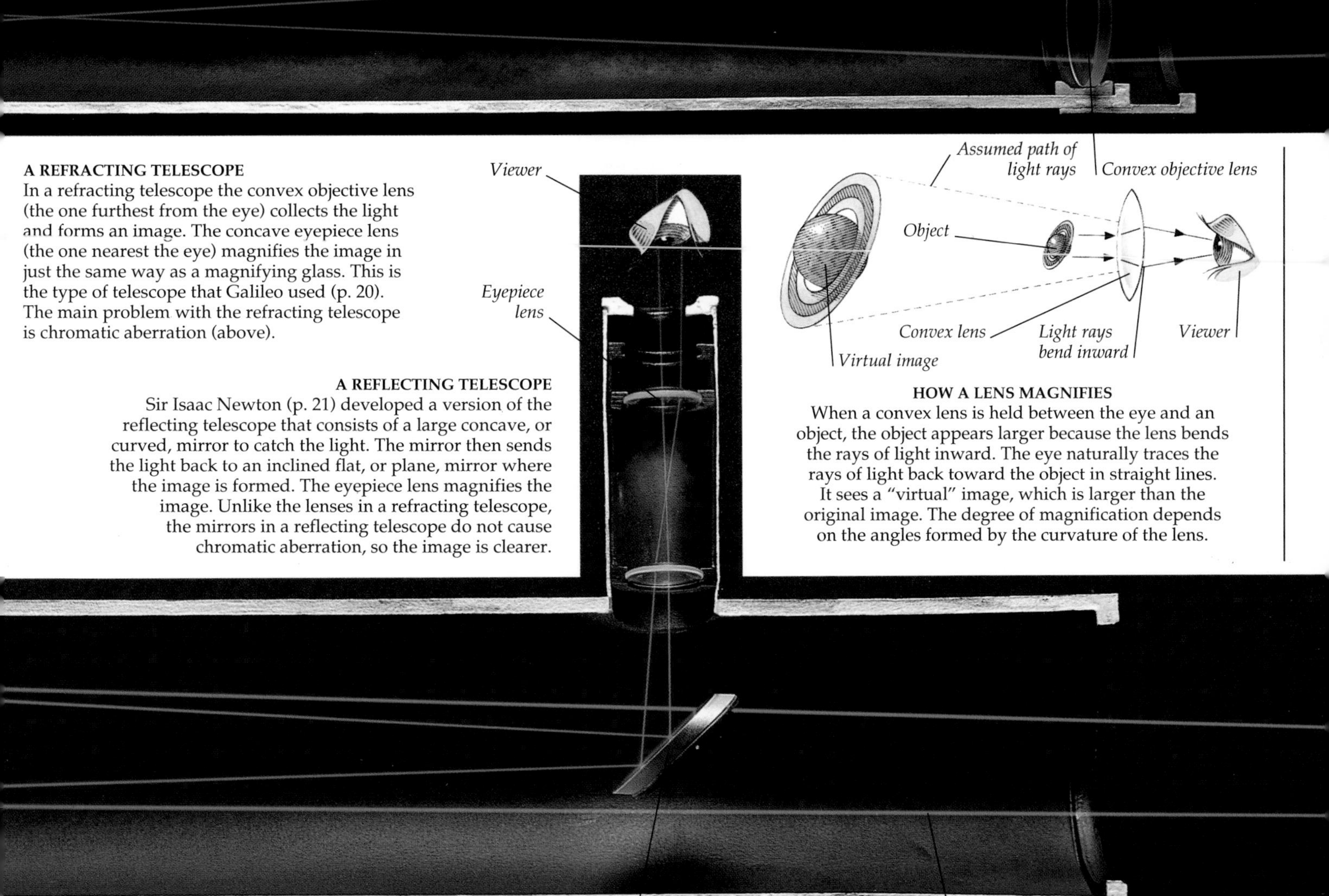

A REFRACTING TELESCOPE
In a refracting telescope the convex objective lens (the one furthest from the eye) collects the light and forms an image. The concave eyepiece lens (the one nearest the eye) magnifies the image in just the same way as a magnifying glass. This is the type of telescope that Galileo used (p. 20). The main problem with the refracting telescope is chromatic aberration (above).

A REFLECTING TELESCOPE
Sir Isaac Newton (p. 21) developed a version of the reflecting telescope that consists of a large concave, or curved, mirror to catch the light. The mirror then sends the light back to an inclined flat, or plane, mirror where the image is formed. The eyepiece lens magnifies the image. Unlike the lenses in a refracting telescope, the mirrors in a reflecting telescope do not cause chromatic aberration, so the image is clearer.

HOW A LENS MAGNIFIES
When a convex lens is held between the eye and an object, the object appears larger because the lens bends the rays of light inward. The eye naturally traces the rays of light back toward the object in straight lines. It sees a "virtual" image, which is larger than the original image. The degree of magnification depends on the angles formed by the curvature of the lens.

The optical telescope

THE MORE LIGHT THAT REACHES THE EYEPIECE in a telescope, the brighter the image of the heavens will be. Astronomers made their lenses and mirrors bigger, they changed the focal length of the telescopes, and combined honeycombs of smaller mirrors to make a single, large reflective surface in order to capture the greatest amount of light and focus it on to a single point. During the 19th century, refracting telescopes (pp. 22-23) were preferred and opticians devoted themselves to perfecting large lenses free of blemishes. In the 20th century there were advances in the technology for grinding and polishing mirrors. A large mirror will intercept more light than a small one, but mirrors larger than 13 ft (4 m) in diameter will sag under their own weight and cause distortion. The development of multiple-mirror instruments in the late 20th century has introduced bigger and better optical telescopes. Smaller mirrors mounted side by side make up a mirror that is bigger than any single one could be.

CAMERAS ON TELESCOPES
Since the 19th century, astronomical photography has been an important tool for astronomers. By attaching a camera to a telescope that has been specially adapted with a motor that can be set to keep the telescope turning at the same speed as the rotation of the Earth, the astronomer can take very long exposures of distant stars (p. 12). Before the invention of photography, astronomers had to draw everything they saw. They had to be artists as well as scientists.

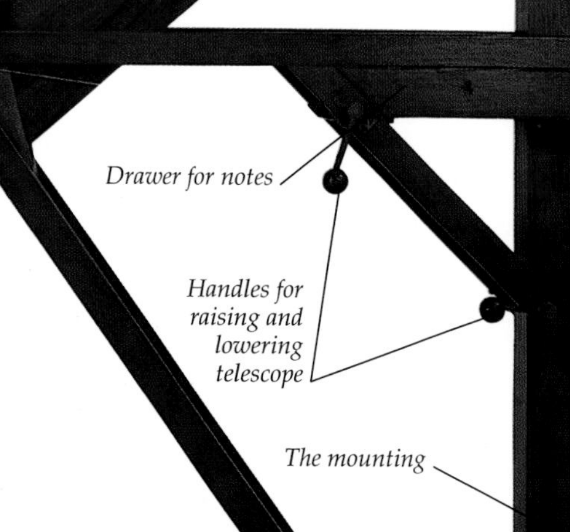

HERSCHEL'S TELESCOPE
First out of economic necessity and later as an indication of his perfectionism, the English astronomer William Herschel (1738-1822) always built his telescopes and hand-ground his own lenses and mirrors. The magnification of a telescope like his 7-ft (2.1-m) Newtonian reflector is about 200 times. This wooden telescope is the kind he would have used during his great survey of the sky, during which he discovered the planet Uranus (pp. 54-55).

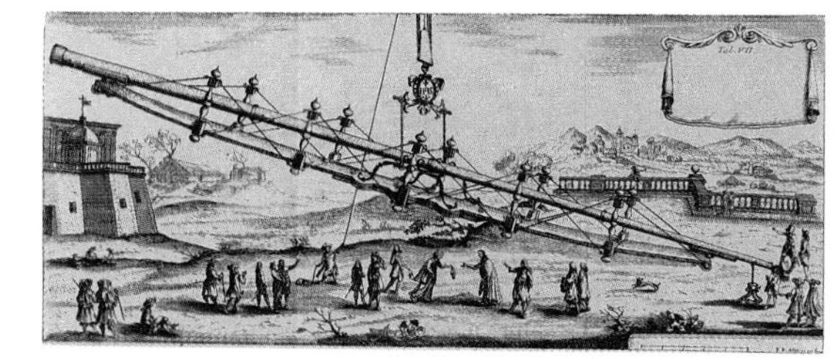

MORE MAGNIFICATION
Increasing the magnification of telescopes was one of the major challenges facing early astronomers. Since the technology to make large lenses was not sufficiently developed, the only answer was to make telescopes with a very long distance between the eyepiece lens and the objective lens. In some instances, this led to telescopes of ridiculous proportions, as shown in this 18th-century engraving. These long focal-length telescopes were impossible to use. The slightest vibration caused by someone walking past would make the telescope tremble so violently that observations were impossible.

AN EQUATORIAL MOUNT
Telescopes have to be mounted in some way. The equatorial mount used to be the favored mount, and is still preferred by amateur astronomers. One axis of the telescope is lined up with the Pole Star (pp. 12-13) which appears stationary. In the southern hemisphere, another bright star is used. The telescope can swing around this axis, automatically following the tracks of stars in the sky as they circle around the Pole Star. The equatorial mount was used for this 28 in (71 cm) refractor, installed at Greenwich, England, in 1893.

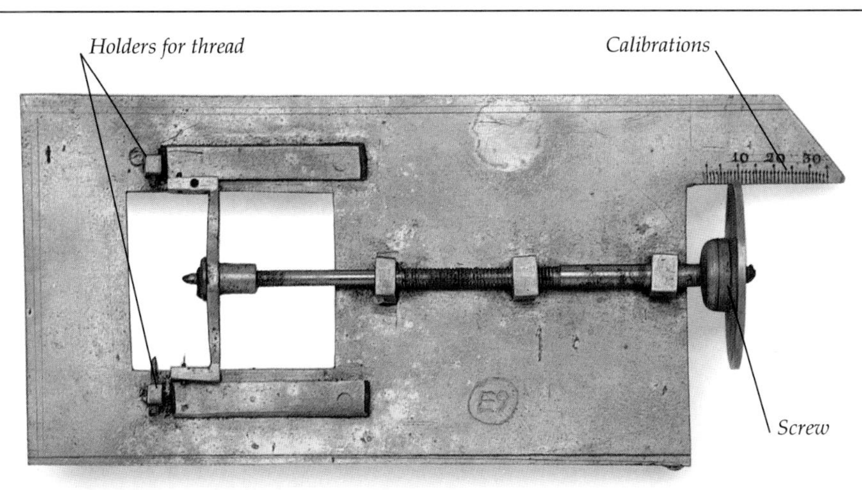

MEASURING ACROSS VAST DISTANCES
The bigger the telescope, the larger its scale will be. This means that measurements become increasingly crude. A micrometer can be set to provide extremely fine gradations, a necessary element when measuring the distances between two stars in the sky that are a very long way away. This micrometer was made by William Herschel. To pinpoint the location of a star, a fine hair or spiderweb was threaded between two holders that were adjusted by means of the finely turned screw on the side.

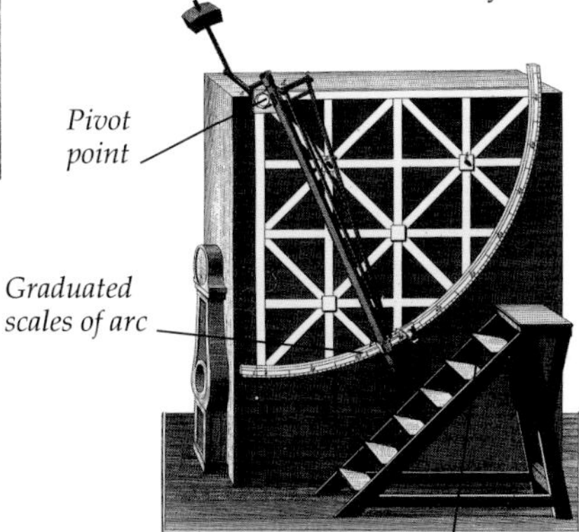

ASTRONOMICAL QUADRANT
Most early telescopes were mounted on astronomical quadrants (p. 12), and to stabilize the telescope, the quadrant was usually mounted on a wall. These kinds of telescopes are called mural quadrants from the Latin word for "wall," *murus*. The telescope was hung on a single pivot-point, so that its eyepiece could be moved along the graduated scale of the arc of the quadrant (p. 12). In this way, astronomers could accurately measure the altitude of the stars they were observing.

AN ALTAZIMUTH MOUNT
The altazimuth mount is like a gun turret. The telescope can track a star by moving up and down as well as turning. It was not a versatile mount for following stars across the sky because stars travel on tilted paths (because of the tilt of the Earth's axis). Today, computers can make the continual adjustments when tracking. Astronomers prefer this mount because it is stable for big telescopes. This is one of the world's largest optical telescopes, located 6,890 ft (2,100 m) above sea level in the Caucasus Mountains, Russia.

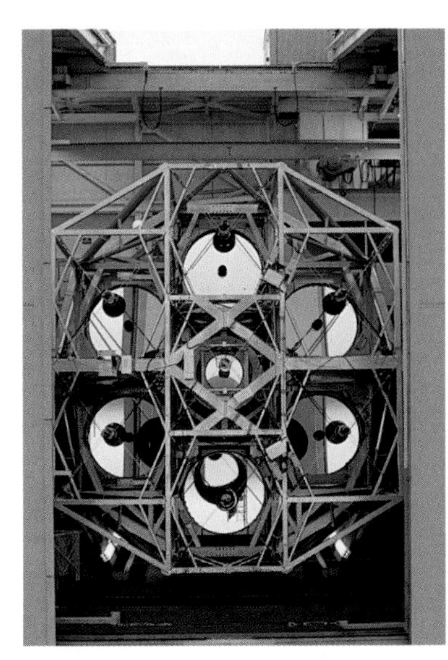

GRINDING MIRRORS
The 16-ft (5-m) mirror of the famous Hale telescope on Mount Palomar in California was cast in 1934 from 35 tons of molten Pyrex. The grinding of the mirror to achieve the correct curved shape was interrupted by World War II. It was not completed until 1947. Mount Palomar was one of the first high-altitude observatories, built where the atmosphere is thinner and the effects of pollution are reduced.

MULTI-MIRROR TECHNOLOGY
The limitations of size imposed by the difficulties of casting a single large mirror have been overcome by the invention of multi-mirror telescope technology (MMT). This telescope in Arizona consists of six separate mirrors, each one being 6 ft (1.8 m) in diameter. These mirrors create a surface equaling 15 ft (4.5 m) in diameter.

Observatories

AN OBSERVATORY IS A PLACE where astronomers watch the heavens. The shapes of observatories have changed greatly over the ages (p. 6). The earliest were quiet places set atop city walls or in towers. Height was important so that the astronomer could have a panoramic, 360° view of the horizon. The Babylonians and the Greeks certainly had rudimentary observatories, but the greatest of the early observatories were those in Islamic North Africa and the Middle East – Baghdad, Cairo, and Damascus. The great observatory at Baghdad had a huge 20-ft (6-m) quadrant and a 56-ft (17-m) stone sextant. It must have looked very much like the observatory at Jaipur – the only one of this type of observatory to remain relatively intact (below). As the great Islamic empires waned and science reawakened in western Europe, observatories took on a different shape. The oldest observatory still in use is the Observatoire de Paris, founded in 1667 (p. 28). A less hospitable climate meant that open-air observatories were impractical. The astronomer and the instruments needed a roof over their heads. Initially, these roofs were constructed with sliding panels or doors that could be pulled back to open the building to the night sky. Since the 19th century, most large telescopes are covered with huge rotatable domes. The earliest domes were made of papier mâché, the only substance known to be sufficiently light and strong. Now most domes are made of fiberglass or high-impact plastics.

THE LEVIATHAN OF PARSONSTOWN
William Parsons (1800-1867), the third Earl of Rosse, was determined to build the largest reflecting telescope. At Parsonstown in Ireland he managed to cast a 72-in (182-cm) mirror, weighing nearly 4 tons and magnifying 800-1,000 times. When the "Leviathan" was built in 1845, it was used by Parsons to make significant discoveries concerning the structure of galaxies and nebulae (pp. 60-63).

BEIJING OBSERVATORY
The Great Observatory set on the walls of the Forbidden City in Beijing, China, was constructed with the help of Jesuit priests from Portugal in 1660 on the site of an older observatory. The instruments included two great armillary spheres (p. 11), a huge celestial globe (p. 10), a graduated azimuth horizon ring, and an astronomical quadrant and sextant (p. 12). The shapes of these instruments were copied from woodcut illustrations in Tycho Brahe's *Mechanica* of 1598 (p. 18).

JAIPUR, INDIA
Early observations were carried out by the naked eye from the top of monumental architectural structures. The observatory at Jaipur in Rajahstan, India, was built by Maharajah Jai Singh in 1726. The monuments include a massive sundial, the Samrat Yantra, and a gnomon inclined at 27°, showing the latitude of Jaipur and the height of the Pole Star (p. 13). There is also a large astronomical sextant and a meridian chamber.

MAUNA KEA
Increasing use of artificial light and air pollution from the world's populous cities have driven astronomers to the most uninhabited regions of the Earth to build their observatories. The best places are mountain tops or deserts in temperate climates where the air is dry, stable and without clouds. The Mauna Kea volcano on the island of Hawaii has the thinner air of high altitudes and the temperate climate of the Pacific. The observatory built there has both optical and radio telescopes (pp. 32-33).

COMPUTER-DRIVEN TELESCOPE
Telescopes have become so big that astronomers are dwarfed by them. This 20-in (51-cm) solar coronagraph in the Crimean Astrophysical Observatory in the Ukraine is driven by computer-monitored engines. A coronagraph is a type of solar telescope that measures the outermost layers of the Sun's atmosphere (p. 38).

What is a meridian?

Meridian lines are imaginary coordinates running from pole to pole that are used to measure distances east and west on the Earth's surface and in the heavens. Meridian lines are also known as lines of longitude. The word meridian comes from the Latin word *meridies*, meaning "the mid-day," because the Sun crosses a local meridian at noon. Certain meridians became important because astronomers used them in observatories when they set up their telescopes for positional astronomy. This means that all their measurements of the sky and the Earth were made relative to their local meridian. Until the end of the 19th century, there were a number of national meridians in observatories in Paris, Cadiz, and Naples.

Prime meridian

THE GREENWICH MERIDIAN
In 1884 there was an international conference in Washington to agree a single Zero Meridian, or Prime Meridian for the world. The meridian running through the Airy Transit Circle – a telescope mounted so that it rotated in a north/south plane – at the Royal Greenwich Observatory outside London was chosen. This choice was largely a matter of convenience. Most of the shipping charts and all of the American railway system used Greenwich as their longitude zero at the time. South of Greenwich the Prime Meridian crosses through France and Africa and then runs across the Atlantic Ocean all the way to the South Pole.

CROSSING THE MERIDIAN
In 1850 the 7th Astronomer Royal of Great Britain, Sir George Biddle Airy (1801-1892), decided he wanted a new telescope. In building it, he moved the previous Prime Meridian for England 19 ft (5.75 m) to the east. The Greenwich Meridian is marked by an illuminated line which bisects Airy's Transit Circle at the Old Royal Observatory (now a museum).

The astronomer

THE MAIN DIFFERENCE BETWEEN ASTRONOMERS and other scientists is that astronomers are not in a position to make things happen. The astronomer has to wait and watch – the key to all astronomy is observing, whether through an optical telescope or by scanning a digitized readout from a computer. There are many different kinds of astronomers. Positional astronomers are mostly interested in measuring the relative positions of the stars. They make the two-dimensional maps of the heavens. Astrophysicists are astronomers who try to understand the physics of the Universe and how matter behaves in the extremes of space. But the types of questions asked by astrophysicists can be traced directly back to the questions of the earliest Greek philosophers – what is the Universe, how is it shaped, and how do I fit into it?

FASHIONABLE AMATEURS
By the 18th century, the science of the stars became an acceptable pastime for the rich and sophisticated. The large number of small telescopes that survive from this period are evidence of how popular amateur astronomy became.

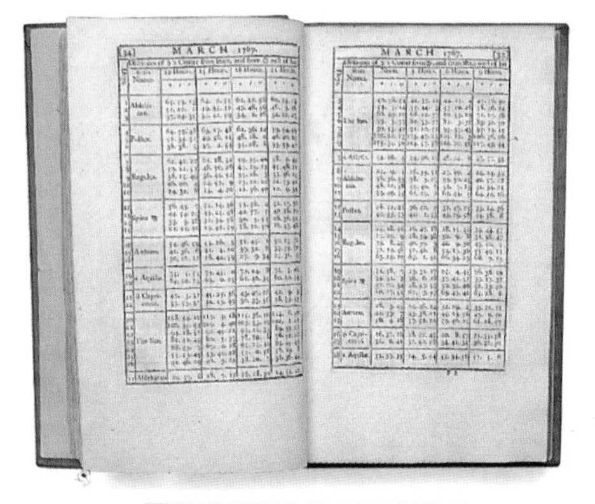

THE NAUTICAL ALMANAC
First published in 1766, *The Nautical Almanac* provides a series of tables showing the distances between certain key stars and the Moon at three-hourly intervals. Navigators can use the tables to help calculate their longitude at sea, when they are out of sight of land (p. 27).

FIRST ASTRONOMER ROYAL
England appointed its first Astronomer Royal, John Flamsteed (1646-1719), in 1675. He lived and worked at the Royal Observatory, Greenwich, built by King Charles II of England in the same year.

IN THE FAMILY
When the Observatoire de Paris was founded in 1667, the French King called a well-known Bolognese astronomer, Gian Domenico Cassini (1625-1712), to Paris to be the observatory's director. He was followed by three generations of Cassinis in the position: Jacques Cassini (1677-1756); César-François Cassini de Thury (1714-1784), who produced the first modern map of France; and Jean-Dominique Cassini (1748-1845). Most historians refer to this great succession of astronomers simply as Cassini I, Cassini II, Cassini III, and Cassini IV.

STAR CLOCK (1815)
One of the primary aspects of positional astronomy is measuring a star's position against a clock. This ingenious clock has the major stars inscribed on the surface of its rotating face. Placing pegs in the holes near the stars to be observed causes the clock to chime when the star is due to pass the local meridian.

ASTRONOMY IN RUSSIA
The Russian astronomer Mikhail Lomonosov (1711-1765) was primarily interested in problems relating to the art of navigation and in fixing latitude and longitude. During his observations of the 1761 Transit of Venus (pp. 46-47), he noticed that the planet seemed "smudgy" and suggested that Venus had a thick atmosphere, many times denser than that of the Earth.

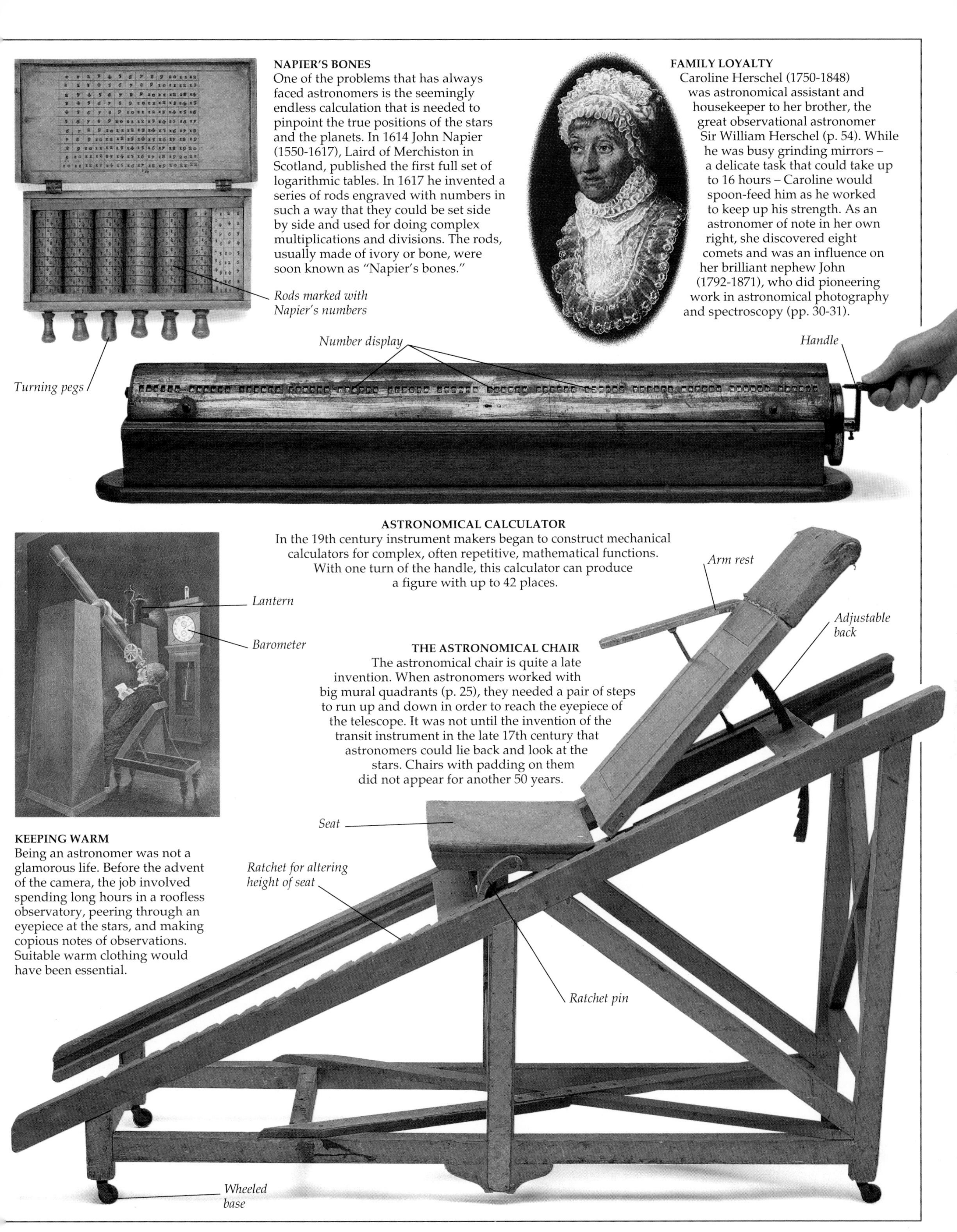

NAPIER'S BONES

One of the problems that has always faced astronomers is the seemingly endless calculation that is needed to pinpoint the true positions of the stars and the planets. In 1614 John Napier (1550-1617), Laird of Merchiston in Scotland, published the first full set of logarithmic tables. In 1617 he invented a series of rods engraved with numbers in such a way that they could be set side by side and used for doing complex multiplications and divisions. The rods, usually made of ivory or bone, were soon known as "Napier's bones."

FAMILY LOYALTY

Caroline Herschel (1750-1848) was astronomical assistant and housekeeper to her brother, the great observational astronomer Sir William Herschel (p. 54). While he was busy grinding mirrors – a delicate task that could take up to 16 hours – Caroline would spoon-feed him as he worked to keep up his strength. As an astronomer of note in her own right, she discovered eight comets and was an influence on her brilliant nephew John (1792-1871), who did pioneering work in astronomical photography and spectroscopy (pp. 30-31).

ASTRONOMICAL CALCULATOR

In the 19th century instrument makers began to construct mechanical calculators for complex, often repetitive, mathematical functions. With one turn of the handle, this calculator can produce a figure with up to 42 places.

KEEPING WARM

Being an astronomer was not a glamorous life. Before the advent of the camera, the job involved spending long hours in a roofless observatory, peering through an eyepiece at the stars, and making copious notes of observations. Suitable warm clothing would have been essential.

THE ASTRONOMICAL CHAIR

The astronomical chair is quite a late invention. When astronomers worked with big mural quadrants (p. 25), they needed a pair of steps to run up and down in order to reach the eyepiece of the telescope. It was not until the invention of the transit instrument in the late 17th century that astronomers could lie back and look at the stars. Chairs with padding on them did not appear for another 50 years.

Spectroscopy

ASTRONOMERS HAVE BEEN ABLE to study the chemical composition of the stars and how hot they are for more than a century by means of spectroscopy. A spectroscope breaks down the "white" light coming from a celestial body into an extremely detailed spectrum. Working on Isaac Newton's discovery of the spectrum (p. 21), a German optician, Josef Fraunhofer (1787-1826), examined the spectrum created by light coming from the Sun and noticed a number of dark lines crossing it. In 1859 another German, Gustav Kirchhoff (1824-1887), discovered the significance of Fraunhofer's lines. They are produced by chemicals in the cooler, upper layers of the Sun (or a star) absorbing light. Each chemical has its own pattern of lines, like a fingerprint. By looking at the spectrum of the Sun, astronomers have found all the elements that are known on the Earth in the Sun's atmosphere.

THE COLORS OF THE RAINBOW
A rainbow is formed by the Sun shining through raindrops against a darker, cloudy background. The light is refracted by droplets of water as if each one were a prism.

Prism splits the light into its colors
The spectrum
Infrared band
Red
Rays of white light
Violet

The spectroscope would be mounted on a telescope here

HERSCHEL DISCOVERS INFRARED
In 1800 Sir William Herschel (p. 54) set up a number of experiments to test the relationship between heat and light. He repeated Newton's experiment of splitting white light into a spectrum (p. 21) and, by masking all the colors but one, was able to measure the individual temperatures of each color in the spectrum. He discovered that the red end of the spectrum is hotter than the violet end, but was surprised to note that an area where he could see no color, next to the red end of the spectrum, is much hotter than the rest of the spectrum. He called this area infrared, or "below the red."

Stand for photographic plate

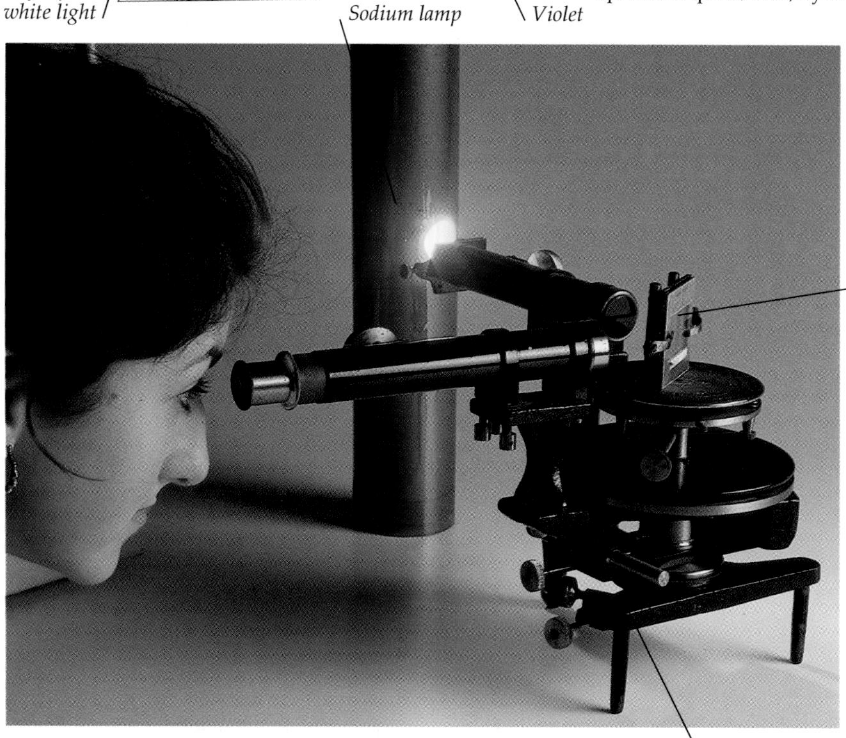
Sodium lamp
Diffraction grating
Spectroscope

LOOKING AT SODIUM
Viewing burning sodium in a flame through a spectroscope can help explain how spectroscopy works in space. According to Gustav Kirchhoff's first law of spectral analysis, a hot dense gas at high pressure produces a continuous spectrum of all colors. His second law states that a hot rarefied gas at low pressure produces an emission line spectrum, that is, a spectrum of bright lines against a dark background. His third law states that when light from a hot dense gas passes through a cooler gas before it is viewed, it produces an absorption line spectrum – a bright spectrum riddled with a number of dark, fine lines.

Solar spectrum showing absorption lines

Sodium

Emission spectrum of sodium *Sodium*

WHAT IS IN THE SUN?
When a sodium flame is viewed through a spectroscope (left), the emission spectrum produces the characteristic bright yellow lines (above). The section of the Sun's spectrum (top) shows a number of tiny "gaps" or dark lines. These are the Fraunhofer lines from which the chemical composition of the Sun can be determined. The two dark lines in the yellow part of the spectrum correspond to the sodium. As there is no sodium in the Earth's atmosphere, it must be coming from the Sun.

KIRCHHOFF AND BUNSEN
Following the invention of the clean-flame burner by the German chemist Robert Bunsen (1811-1899), it was possible to study the effect of different chemical vapors on the known pattern of spectral lines. Together, Gustav Kirchhoff and Bunsen invented a new instrument called the spectroscope to measure these effects. Within a few years, they had managed to isolate the spectra for many known substances, as well as discovering a few unknown elements.

Continuous spectrum

ABSORBING COLOR
To prove his laws of spectral analysis, Kirchhoff used sodium gas to show that when white light is directed through the gas, the characteristic color of the sodium is absorbed and the spectrum shows black lines where the sodium should have appeared. In the experiment shown above, a continuous spectrum (top) is produced by shining white light through a lens. When a petri dish of the chemical potassium permanganate in solution is placed between the lens and the light, some of the color of the spectrum is absorbed.

The spectrum of potassium permanganate

SPECTRUM OF THE STARS
By closely examining the spectral lines in the light received from a distant star or planet, the astronomer can detect these "fingerprints" and uncover the chemical composition of the object being viewed. Furthermore, the heat of the source can be discovered by studying the spectral lines. Temperature can be measured by the intensities of individual lines in their spectra. The width of the line provides information about temperature, movement, and presence of magnetic fields. With magnification, each of these spectra can be analyzed in more detail.

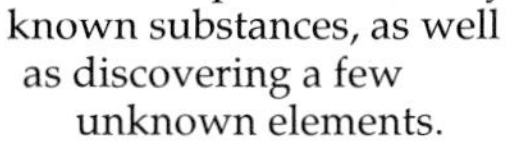

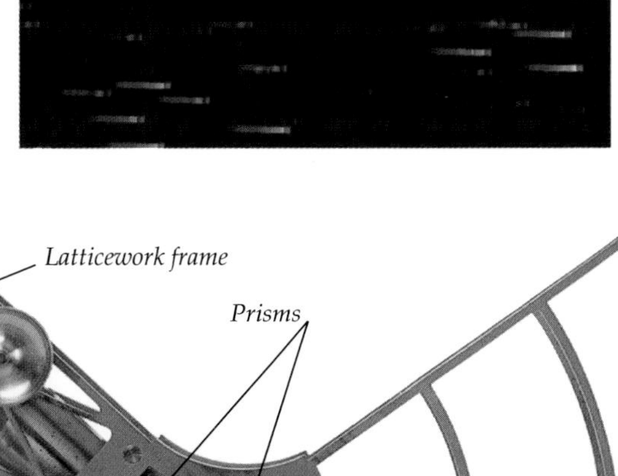

Eyepiece

Latticework frame

Prisms

Micrometer (p. 25)

Eyepiece

NORMAN LOCKYER (1836-1920)
During the solar eclipse of 1868, a number of astronomers picked up a new spectral line in the upper surface of the Sun, the chromosphere (p. 39). The English astronomer Lockyer realized that the line did not coincide with any of the known elements. The newly discovered element was named helium (*helios* is Greek for the sun god). Finally, in 1895, helium was found to be present on the Earth.

THE SPECTROSCOPE
A spectroscope uses a series of prisms or a diffraction grating – a device that diffracts light through fine lines to form a spectrum – to split light into its constituent wavelengths (pp. 32-33). Before the era of photography, an astronomer would view the spectrum produced with the eye, but now it is mostly recorded with an electronic detector called a CCD (p. 45). This 19th-century spectroscope uses a prism to split the light.

The radio telescope

WITH THE DISCOVERY OF non-visible light, such as infrared (p. 30), and electromagnetic and X-ray radiations, scientists began to wonder if objects in space might emit invisible radiation as well. The first such radiation to be discovered (by accident) was radio waves – the longest wavelengths of the electromagnetic spectrum. To detect radio waves, astronomers constructed huge dishes in order to capture the long waves and "see" detail. Even so, early radio telescopes were never large enough, proportionally, to catch the fine features that optical telescopes could resolve. Today, by electronically combining the output from many radio telescopes, a dish the size of the Earth can be synthesized, revealing details many times finer than optical telescopes. Astronomers routinely study all radiations from objects in space, often using detectors high above the Earth's atmosphere (p. 7).

ANDERS ÅNGSTRÖM (1814-1874)
Ångström was a Swedish astronomer who mapped the Fraunhofer lines in the Sun's spectrum (p. 30). The units needed to measure the wavelengths of the colors corresponding to these lines were so short that Ångström had to invent a new unit of length. The Ångström (symbol Å) is 10^{-10} meters.

EVIDENCE OF RADIO RADIATION
The first evidence of radio radiation coming from outer space was collected by the American scientist Karl Jansky (1905-1950) who, in 1931, using homemade equipment (above), investigated the static affecting short-wavelength radio-telephone communication. He deduced that this static must be coming from the center of our galaxy (pp. 62-63).

AMATEUR ASTRONOMER
On hearing about Jansky's discoveries, American amateur astronomer Grote Reber (1911-) built a large, moveable radio receiver in his backyard in 1936. It had a parabolic surface to collect the radio waves. With this 29-ft (9-m) dish, he began to map the radio emissions coming from the Milky Way. For years Reber was the only radio astronomer in the world.

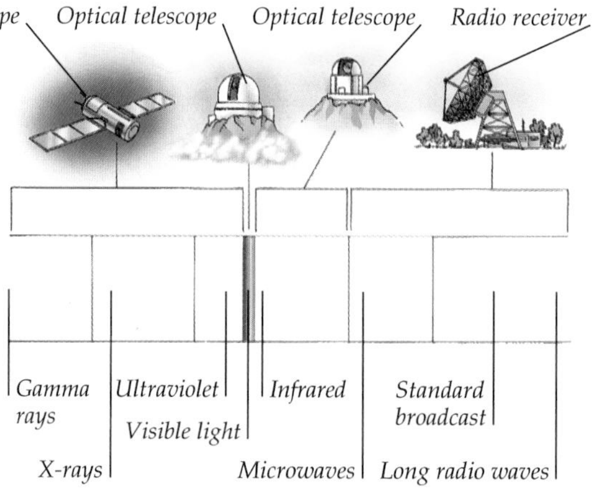

ELECTROMAGNETIC SPECTRUM
The range of frequencies of electromagnetic radiation is known as the electromagnetic spectrum. Very low on the scale are radio waves, rising to infrared (p. 30), visible light, ultraviolet, and X-rays, with gamma rays at the highest frequency end of the spectrum. The radiations that pass through the Earth's atmosphere are light and radio waves, though infrared penetrates to the highest mountaintops. The remainder can only be detected by sending instruments into space in special probes (pp. 34-35). All telescopes – radio, optical, and infrared – "see" different aspects of the sky caused by the different physical processes going on.

ARECIBO RECEIVER
The mammoth Arecibo radio receiver is built in a natural limestone concavity in the jungle south of Arecibo, Puerto Rico. The "dish," which is a huge web of steel mesh, measures 1,000 ft (305 m) across, providing a 50-acre (20-hectare) collecting surface. Although the dish is fixed, overhead antennas can be moved to different parts of the sky.

HOT SPOTS
Radio astronomers can create temperature maps of planets. This false-color map shows temperatures just below Mercury's surface. Because Mercury is so close to the Sun, the hottest area is on Mercury's equator, shown here as red. The blue areas are the coolest.

BERNARD LOVELL
The English astronomer, Bernard Lovell (1913-), was a pioneer of radio astronomy. He developed a research station at Jodrell Bank, England, in 1945 using surplus army radar equipment. He is seen here reading a screen in the control room of the 250-ft (76-m)-diameter Mark 1 radio telescope. The giant dish, which is partially visible in the background, was commissioned in 1957.

HIGH-TECH TELESCOPE
Communications technology allows astronomers to work nearly anywhere in the world. All they need is a computer linkup. While optical telescopes are sited far from built-up areas (p. 27), clear skies are not necessary for radio astronomy. This telescope is the world's largest, fully steerable, single-dish radio telescope; it is 330 ft (100 m) in diameter and is located near Bonn, Germany.

HOW A RADIO TELESCOPE WORKS
The parabolic dish of a radio telescope can be steered to pick up radio signals. It focuses them to a point from which they are sent to a receiver, a recorder, and then a data room at a control center. Computer equipment then converts intensities of the incoming radio waves into images that are recognizable to our eyes as objects from space (p. 57).

THE VERY LARGE ARRAY
Scientists soon realized that radio telescopes could be connected together by electrical wires to form very large receiving surfaces. For example, two dishes 62 miles (100 km) apart can be linked electronically so that their receiving area is the equivalent of a 62-mile (100-km)-wide dish. One of the largest arrangements of telescopes is the Very Large Array (VLA) set up in the desert near Socorro, New Mexico. Twenty-seven parabolic dishes have been arranged in a huge "Y," covering more than 17 miles (27 km).

Venturing into space

Luna 1

SKEPTICS PROPHESIED AN END to space exploration after the *Challenger* Space Shuttle was destroyed in 1986 and the Hubble space telescope had trouble with its mirror in 1993. With the end of the tensions of the Cold War in 1989, many people could see no point in spending vast sums of money to support competitive space programs when economies on the Earth were under strain. However, the technology underpinning the space race has brought improvements to our lives. The development of new, lightweight, yet strong, materials and water purification methods are direct spin-offs of space research. Most of the long-distance telecommunications on the Earth now rely on orbiting satellites. Navstars are satellites used by ships and planes for navigational purposes. Military satellites are used in early-warning systems. Our weather forecasts – which come from meteorological satellites – are now more accurate, while resource satellites monitor the Earth's surface.

LUNAR PROBES
The former USSR launched *Sputnik 1*, the first artificial satellite, into space in 1957. Between the late 1950s and 1976, several probes were sent to explore the surface of the Moon. *Luna 1* was the first successful lunar probe. It passed within 3,730 miles (6,000 km) of the Moon. *Luna 3* was the first probe to send back pictures to the Earth of the far side of the Moon (pp. 40-41). The first to achieve a soft landing was *Luna 9* in February 1966. *Luna 16* collected soil samples, bringing them back without any human involvement. The success of these missions forced people to take space exploration more seriously.

GETTING INTO SPACE
The American physicist Robert Goddard (1882-1945) launched the first liquid-fueled rocket in 1926. This fuel system overcame the major obstacle to launching an orbiting satellite, which was the weight of solid fuels. If a rocket is to reach a speed great enough to escape the Earth's gravitational field, it needs a thrust greater than the weight it is carrying.

THE FIRST MAN IN SPACE
On April 12, 1961 the USSR launched the 5-ton spaceship *Vostok 1*. It was manned by the cosmonaut Yuri Gagarin (1934-1968), who made a complete circuit of the Earth at a height of 188 miles (303 km). He remained in space for 1 hour and 29 minutes before landing back safely in the USSR. He was hailed as a national hero and is seen here being lauded by the Premier of the USSR, Nikita Khruschev.

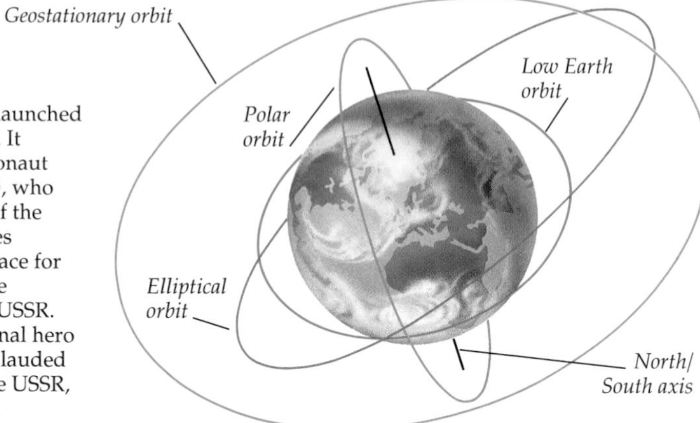

SATELLITE ORBITS
A satellite is sent into an orbit that is most suitable for the kind of work it has to do. Space telescopes such as Hubble (p. 7), take the low orbits – 186 miles (300 km) above the Earth's surface. U.S. spy and surveillance satellites orbit on a north/south axis to get a view of the whole Earth, while those belonging to Russia often follow elliptical orbits that allow them to spend more time over their own territory. Communications and weather satellites are positioned above the equator. They take exactly 24 hours to complete an orbit, and therefore seem to hover above the same point on the Earth's surface – known as a geostationary orbit.

LUNAR LANDING
Between 1969 and 1972, six manned lunar landings took place. The first astronaut to set foot on the Moon was Neil Armstrong (1930-) on July 20, 1969. Scientifically, one of the major reasons for Moon landings was to try to understand the origin of the Moon itself and to understand its history and evolution. This photograph shows American astronaut James Irwin with the *Apollo 15* Lunar Rover in 1971

The Space Shuttle

Launching satellites by rocket is expensive and wasteful because the launch vehicle is destroyed in the process. The U.S. Space Shuttle has proved to be a reusable alternative. Its missions last about a week and have included taking astronauts to the Hubble space telescope for repairs and carrying modules into space for the construction of a new space station.

COOPERATION IN SPACE
The European Space Agency's rocket, *Ariane*, is used by a number of European countries to launch a series of communications satellites. The satellite or probe that is carried into space is known as a payload. The idea of cooperation in space is markedly different from the 1960s and 1970s, when the U.S. and the USSR competed against each other. *Ariane* provides the means by which a group of less wealthy countries can band together and share the benefits of space-age technology. This photograph shows *Ariane 3* taking off from French Guiana in 1984. Unfortunately the rocket crashed shortly after this photograph was taken.

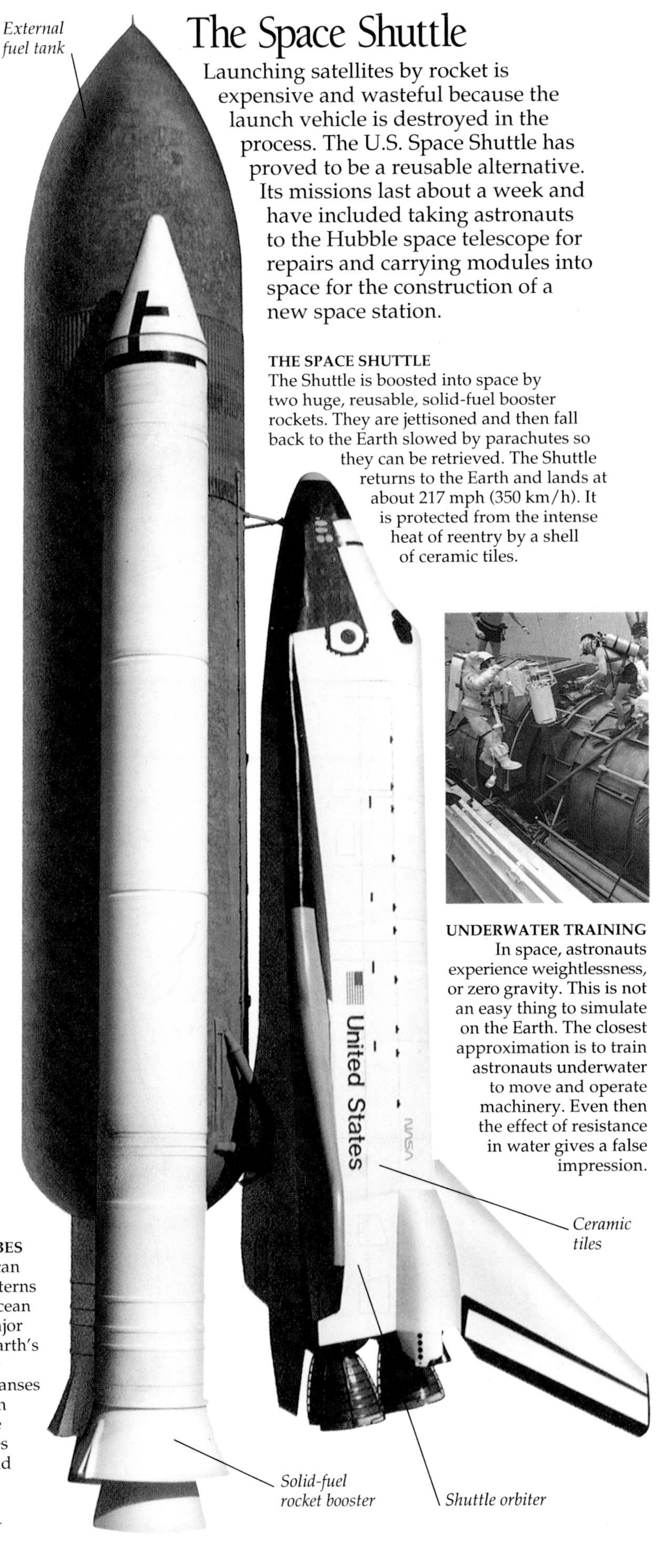

THE SPACE SHUTTLE
The Shuttle is boosted into space by two huge, reusable, solid-fuel booster rockets. They are jettisoned and then fall back to the Earth slowed by parachutes so they can be retrieved. The Shuttle returns to the Earth and lands at about 217 mph (350 km/h). It is protected from the intense heat of reentry by a shell of ceramic tiles.

UNDERWATER TRAINING
In space, astronauts experience weightlessness, or zero gravity. This is not an easy thing to simulate on the Earth. The closest approximation is to train astronauts underwater to move and operate machinery. Even then the effect of resistance in water gives a false impression.

LIVING IN SPACE
The former USSR launched the first module of its *Mir* space station in 1986. The station is a central living quarters surrounded by a series of docking ports. These ports can be used to expand the station by adding more modules, or entry points, for cargo as well as supply ferries from the Earth. Some cosmonauts have lived there for more than a year. Scientists still have a great deal to learn about the effects on humans of prolonged periods in deep space.

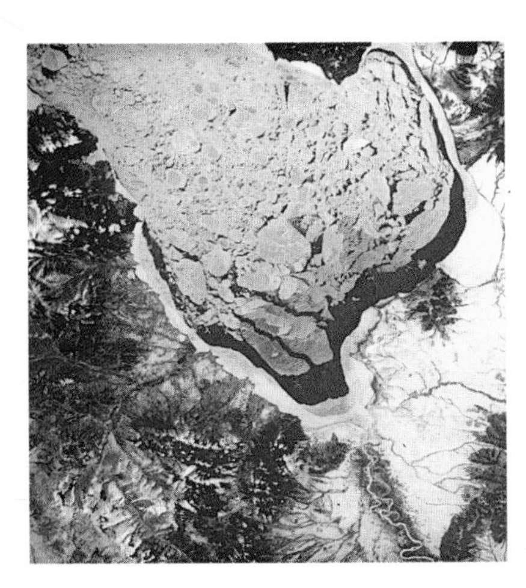

BENEFITS OF SPACE PROBES
Meteorological satellites can monitor the changing patterns of the weather and plot ocean currents, which play a major role in determining the Earth's climate. Data gathered by monitoring such vast expanses as this Russian ice floe can be used to predict climate change. Resource satellites are used for geological and ecological research. For example, they map the distribution of plankton – a major part of the food chain – in ocean waters.

The Solar System

THE SOLAR SYSTEM is the group of planets, moons, and space debris orbiting around our Sun. It is held together by the gravitational pull of the Sun, which is nearly 1,000 times more massive than all the planets put together. The Solar System was probably formed from a huge cloud of interstellar gas and dust that contracted under the force of its own gravity five billion years ago. The planets are divided into two groups. The four planets closest to the Sun are called "terrestrial" from the Latin word *terra* meaning "land" because they are small and dense and have hard surfaces. Four of the outer planets beyond Mars are called "Jovian" because they all resemble Jupiter – Jove was another name for the Roman god, Jupiter. These outer planets are largely made up of gases. Between Mars and Jupiter there is a zone of space debris – the asteroid belt (p. 58).

THE SECRET OF ASTRONOMY
This allegorical engraving shows Astronomy, with her star-covered robe, globe, telescope, and quadrant, next to a female figure who might represent Mathematics. The small angel between them holds a banner proclaiming *pondere et mensura*: "to weigh and measure" – which is the secret of the art of astronomy.

RELATIVE SIZE
The Sun has a diameter of approximately 865,000 miles (1,392,000 km). It is almost 10 times larger than the largest planet, Jupiter, which is itself big enough to contain all the other planets put together. The planets are shown here to scale against the Sun. Those planets with orbits inside the Earth are sometimes referred to as the inferior planets; those beyond the Earth are the superior planets. The four small planets that orbit the Sun relatively closely – Mercury, Venus, Earth, and Mars - have lower masses than those of the outer four, but have much greater densities (p. 45). Jupiter, Saturn, Uranus, and Neptune have large masses with low densities. They are more widely spaced apart and travel at great distances from the Sun. Pluto is the smallest planet in the Solar System and is in a category of its own (p. 57).

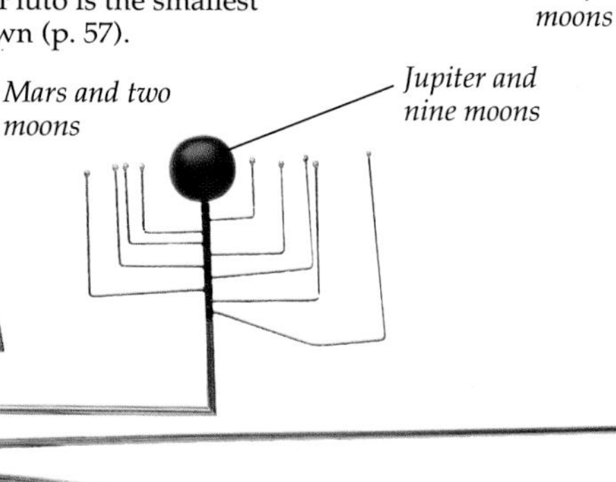
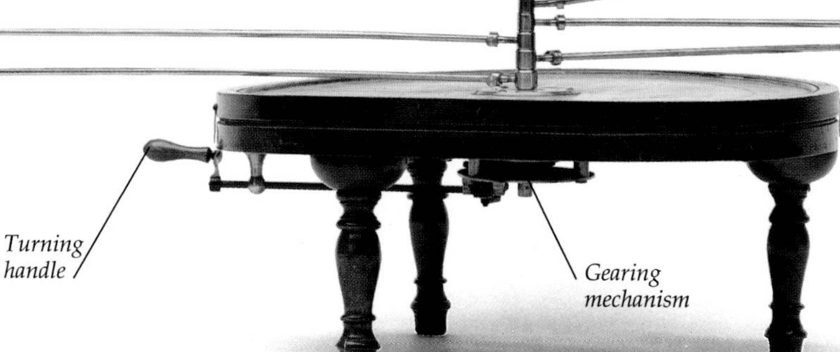

TEACHING ASTRONOMY
During the 19th century, the astronomy of the Solar System was taught by mechanical instruments such as this planetarium. The complex gearing of the machine is operated by a crank handle, which ensures that each planet completes its solar orbit relative to the other planets. The planets are roughly to a scale of 50,000 miles (80,500 km) to 1 in (3 cm), except for the Sun, which would need to be 17 in (43 cm) in diameter for the model to be accurate.

CELESTIAL MECHANICS
The Frenchman Pierre Simon Laplace (1749-1827) was the first scientist to make an attempt to compute all the motions of the Moon and the planets by mathematical means. In his five-volume work, *Traité de méchanique céleste* (1799-1825), Laplace treated all motion in the Solar System as a purely mathematical problem, using his work to support the theory of universal gravitation (p. 21). His idea, for which he was severely criticized during the following century, was that the heavens were a great celestial machine, like a timepiece that, once set in motion, would go on forever.

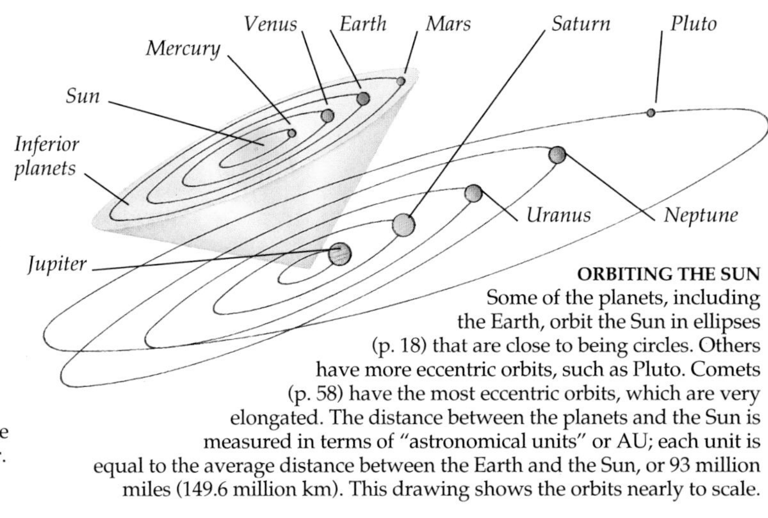

ORBITING THE SUN
Some of the planets, including the Earth, orbit the Sun in ellipses (p. 18) that are close to being circles. Others have more eccentric orbits, such as Pluto. Comets (p. 58) have the most eccentric orbits, which are very elongated. The distance between the planets and the Sun is measured in terms of "astronomical units" or AU; each unit is equal to the average distance between the Earth and the Sun, or 93 million miles (149.6 million km). This drawing shows the orbits nearly to scale.

Photographing the planets

One of the key tasks of space probes (pp. 34-35) is to send pictures back to the Earth that are recognizable. The differing quality of each of the individual "snapshots" is rationalized when the images are scanned into a computer in digital form. One device that does this is the CCD, or charge coupled device (p. 7), a silicon chip divided into 64,000 pixels, or units. Each pixel records a different brightness – the larger the number, the greater the light in the image. The value of each pixel is recorded and fed into a computer. The readout is displayed on a monitor from which pictures or slides can be made.

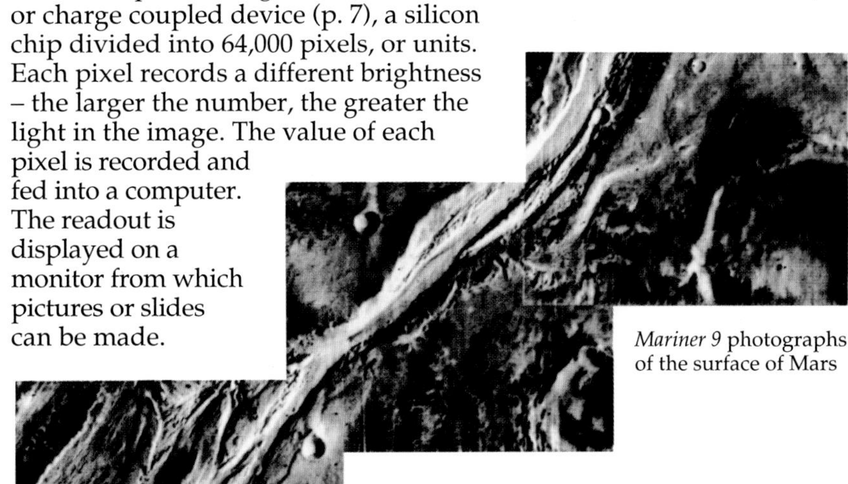

Mariner 9 photographs of the surface of Mars

COLLATING 2-DIMENSIONAL IMAGES
Normal color film is rarely used for photographing planetary bodies because they are low in contrast and because the exposures need to be very long (at least 1 hour). The images are taken in black and white and pieced together like a mosaic.

Escaping elements

Lighter elements

By increasing the vibration, the balls are given more energy

Heavier elements

Kinetic energy machine

HYDROGEN IN THE SOLAR SYSTEM
Hydrogen is a common element in the Solar System. Hydrogen atoms are so energetic that lightweight planets cannot hang onto them. This is why the heavier nitrogen makes up such a high percentage of the Earth's atmosphere (p. 42). Lighter hydrogen has escaped because the Earth's gravity is not strong enough to hold onto it. The red balls in this kinetic energy machine represent the heavier elements; the tiny silver balls represent the lighter elements such as hydrogen. Our massive Sun is made up largely of hydrogen. Its great mass pulls the hydrogen inward and, at its core, hydrogen fuses into helium under the extreme heat and pressure. It is this reaction, like a giant hydrogen bomb, releasing lots of energy, that makes the Sun shine. Hydrogen also makes up a large part of the gas giants (pp. 52-55).

COLOR MOSAIC OF MARS
This mosaic picture of Mars is produced by a computer using 50 images taken by the *Viking* orbiter. The color in the image is made by adding red, green, and violet filters to give an image in full color.

The Sun

ALMOST EVERY ANCIENT CULTURE recognized the Sun as the giver of life and primary power behind events here on the Earth. The Sun is the center of our Solar System, our local star. It has no permanent features because it is gaseous – mainly incandescent hydrogen. Since it is so close, astronomers study the Sun in order to understand the nature of the stars in general. They focus mostly on the layers of atmosphere that extend out into space – the corona, the chromosphere, and the photosphere. By using spectroscopic analysis (pp. 30-31), scientists know that the Sun, like most stars (pp. 60-61), is made up mostly of hydrogen. In its core, the hydrogen nuclei are so compressed that they eventually fuse into helium. This is the same thing that happens in a hydrogen atomic bomb. Every minute, the Sun converts 240 million tons of mass into energy. Albert Einstein's famous formula, $E=mc^2$, shows how mass and energy are mutually interchangeable (p. 63), helping scientists understand the source of the Sun's energy.

THE CORONAGRAPH
In 1930 the French astronomer Bernard Lyot (1897-1952) invented the coronagraph. It allows the Sun's corona to be viewed without waiting for a total solar eclipse.

VIEWING THE SUN
Even though the Sun is more than 92 million miles (149 million km) from the Earth, its rays are still bright enough to damage the eyes permanently. The Sun should *never* be viewed directly, and certainly not through a telescope or binoculars. Galileo went blind looking at the Sun. This astronomer is at the Kitt Peak National Observatory in Arizona. Two mirrors at the top of the solar telescope tower reflect the Sun's image down a tube to the mirror below. Inside the tube there is a vacuum. This prevents distortion that would be caused by the air in the tower.

THE DIPLEIDOSCOPE
Local noon occurs when the Sun crosses the local north/south meridian (p. 27). In the 19th century, a more accurate device than the gnomon (p. 14) was sought to indicate when noon occurred. The dipleidoscope, invented in 1842, is an instrument with a hollow, right-angled prism, which has two silvered sides and one clear side. As the Sun passes directly overhead, the two reflected images are resolved into a single one. This shows when it is local noon.

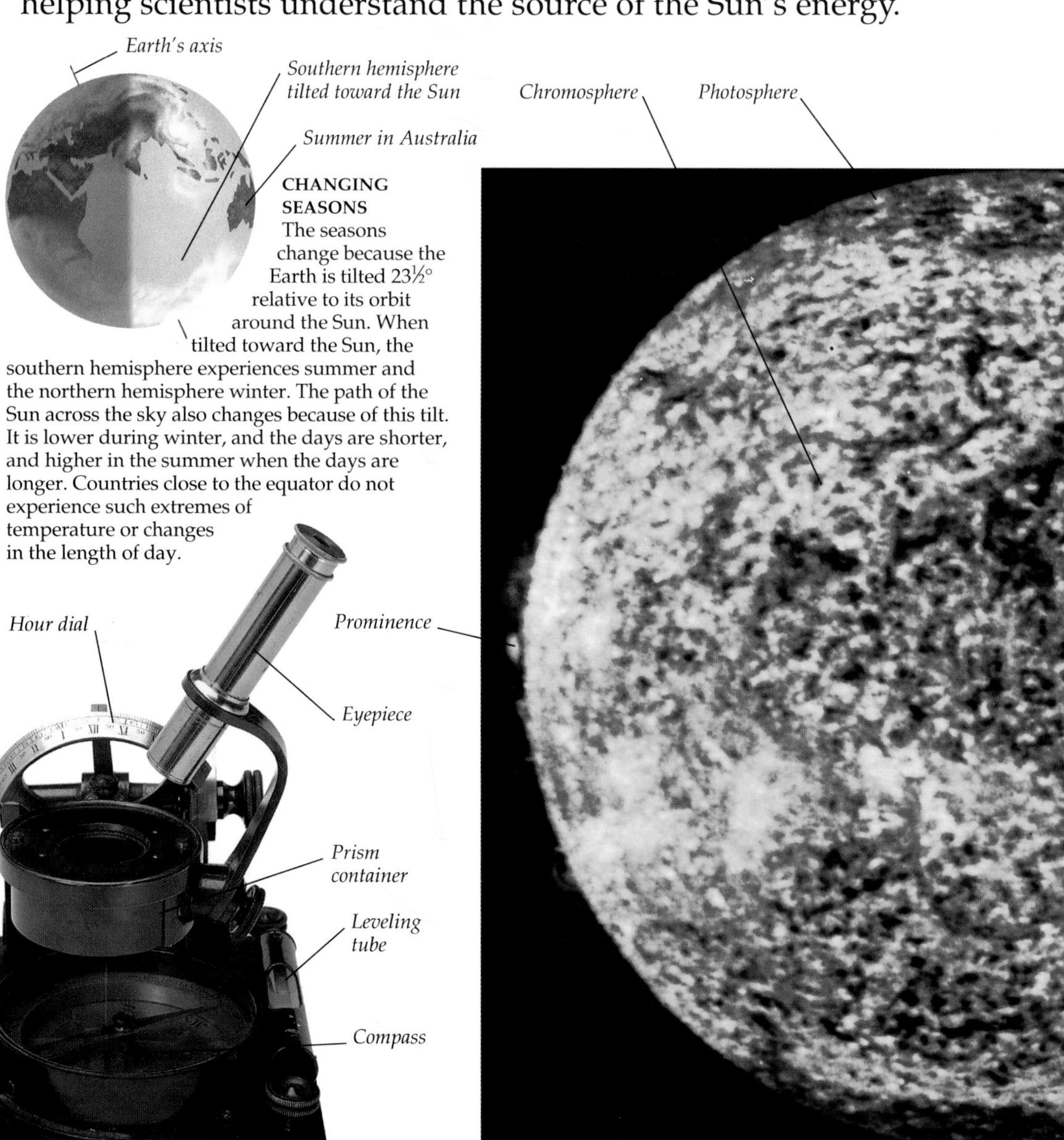

CHANGING SEASONS
The seasons change because the Earth is tilted 23½° relative to its orbit around the Sun. When tilted toward the Sun, the southern hemisphere experiences summer and the northern hemisphere winter. The path of the Sun across the sky also changes because of this tilt. It is lower during winter, and the days are shorter, and higher in the summer when the days are longer. Countries close to the equator do not experience such extremes of temperature or changes in the length of day.

Sunspots

Sunspots are cooler areas on the Sun's surface, where the magnetic field has disturbed the passage of heat from the core to its photosphere. When sunspots are at a maximum, the Sun also experiences large explosive eruptions called flares. These flares blast large numbers of charged particles into space. If these particles reach the Earth, they can cause disruption in radio communications and the wonderful effects of the aurora borealis and aurora australis.

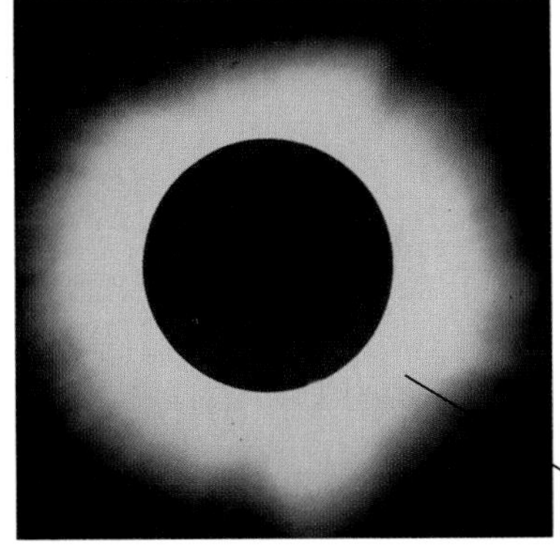

THE CORONA
The outermost layer of the Sun's atmosphere is called the corona. Even though it extends millions of miles into space, it cannot be seen during the day because of the brightness of the blue sky. During a total eclipse, the corona appears like a crown around the Moon. It is clearly seen in this picture of a total eclipse over Mexico in March 1970.

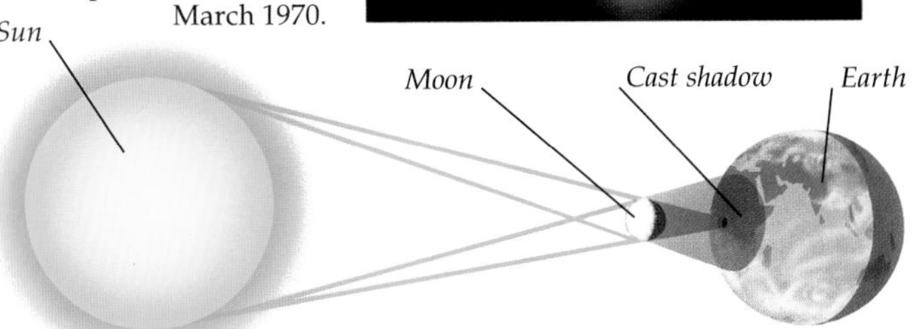

SOLAR ECLIPSE
A solar eclipse happens when a new Moon passes directly between the Earth and the Sun, casting a shadow on the surface of the Earth. From an earthly perspective, it looks as if the Moon has blocked out the light of the Sun. Total eclipses of the Sun are very rare in any given location, occurring roughly once every 360 years in the same place. However, several solar eclipses may occur each year.

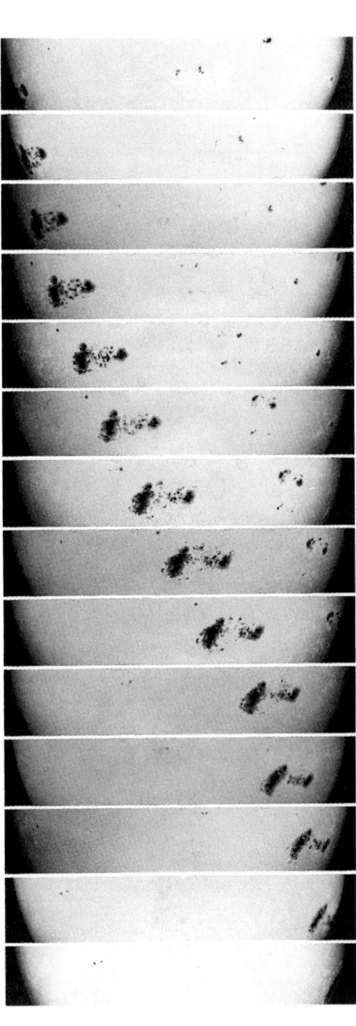

PLOTTING THE SUNSPOTS
Sunspots moving somewhat randomly across the surface of the Sun show that the Sun is actually spinning. Unlike the planets, however, the whole mass of the Sun does not spin at the same rate because it is not solid. The Sun's equator takes 25 Earth days to make one complete rotation. The Sun's poles take nearly 30 days to accomplish the same task. These photographs are a record of the movements of a large spot group on 14 days in March/April 1947.

A PHOTOHELIOGRAPH
Astronomers watch sunspots through an adapted telescope. Instead of an eyepiece, there is a large screen of filtered glass. The Sun's image is projected onto the glass, which is marked with coordinates. The astronomer can count and measure sunspots without looking directly at the Sun.

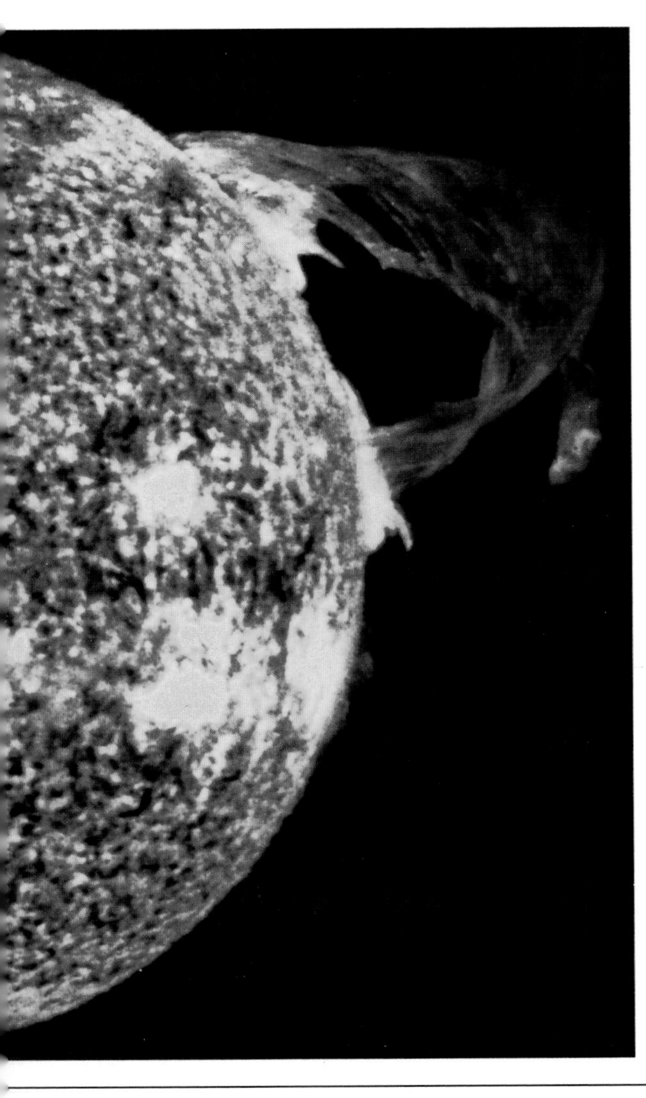

SOLAR ATMOSPHERE
As the Sun is a star, it does not have a surface. It has layers of gases of different densities. Most of the visible light of the Sun comes from the photosphere, which is about 186-248 miles (300-400 km) thick. The next layer out is the chromosphere, which is hotter than the photosphere. The outer layer is called the corona. Since the Sun is in a constantly volatile state, its layers are always subject to disturbance. Prominences and spectacular solar flares – huge eruptions from the corona that extend outwards for tens of thousands of miles – are directly related to disturbances of the Sun's magnetic field.

FACTS ABOUT THE SUN

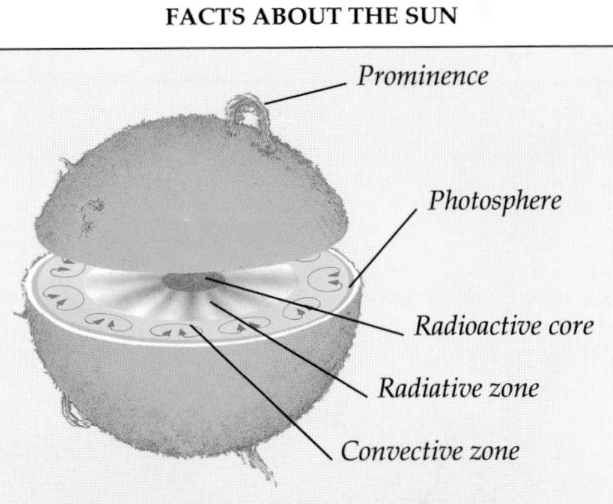

- **Equatorial diameter** 0.86 million miles/1.4 million km
- **Distance from the Earth** 93 million miles/149 million km
- **Rotational period** 25 Earth days
- **Volume** (Earth =1) 1,306,000
- **Mass** (Earth =1) 333,000
- **Density** (water = 1) 1.41
- **Temperature at surface** 10,000°F

The Moon

The Moon is the Earth's only satellite, averaging about 239,000 miles (384,000 km) away. Next to the Sun it is the brightest object in our sky, more than 2,000 times as bright as Venus. The origin of the Moon is still the subject of heated debate. Some scientists believe that the Earth and the Moon were formed at the same time from the dust and gas of the primordial Solar System. Others argue that the Moon was a body passing the Earth that got caught by the Earth's gravitational field. Some think that a body the size of Mars once collided with the Earth. The impact splashed debris into space, some of which eventually formed the Moon.

EARLY MOON MAP
The same side of the Moon always faces toward the Earth. Because the Moon's orbit is not circular and it travels at different speeds, we can see more than just the face of the Moon. This phenomenon, called libration, means that about 59 percent of the Moon's surface is visible from the Earth. In 1647 Johannes Hevelius (1611-1687) published his lunar atlas *Selenographia* showing the Moon's librations.

Shadow is used to calculate the height of crater walls

Floor of the crater

Crater walls

COPERNICUS CRATER
The Moon's craters were formed between 3.5 and 4.5 billion years ago by the impact of countless meteorites. These impact craters are all named after famous astronomers and philosophers. Because the Moon has no atmosphere, there has been little erosion of its surface. This plaster model shows Copernicus crater, which is 56 miles (90 km) across and 11,000 ft (3,352 m) deep. Inside the crater there are mountains with peaks 8 miles (5 km) above the crater's floor.

Sun

Moon's orbit

Earth

Umbra or total shadow

Moon

Penumbra, or partial shadow

A LUNAR ECLIPSE
An eclipse happens when the Earth passes directly between the Sun and the full Moon, so that the Earth's shadow falls on to the surface of the Moon. This obscures the Moon for the duration of the eclipse.

Lunar equator

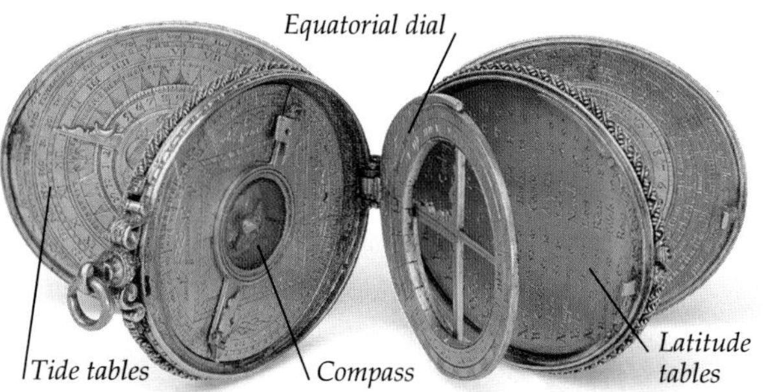

Equatorial dial

Tide tables

Compass

Latitude tables

TIDE TABLES
The pull of the force of gravity of the Moon (p. 21), and to a lesser extent, of the Sun, causes the water of the seas on the Earth to rise and fall. This effect is called a tide. When the Sun, the Moon, and the Earth are all aligned at a New or Full Moon, the tidal "pull" is the greatest. These are called Spring tides. When the Sun and the Moon are at right angles to each other, they produce smaller pulls called Neap tides. This compendium (1569) contains plates with tables indicating the tides of some European cities. It was an essential instrument for sailors entering harbor.

PHASES OF THE MOON
The phases of the Moon are caused by the constantly changing series of angles formed by the Sun and the Moon as the Moon revolves around the Earth. When the Moon and the Sun are on opposite sides of the Earth, the Sun shines directly on the Moon's surface, resulting in a Full Moon. When the area of the lit surface increases, the Moon is said to be waxing; as it decreases, it is said to be waning.

Waxing crescent Moon at 4 days — Full Moon at 14 days — Waning 19-day Moon — Moon at 21 days — Moon at 24 days

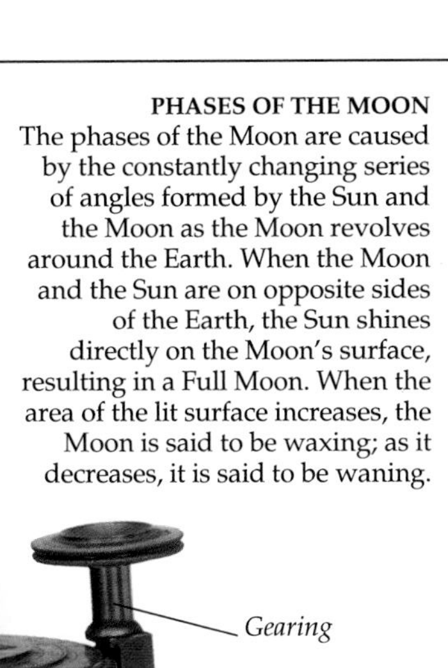

THE SURFACE OF THE MOON
The features on the far side of the Moon were a mystery until the late 1950s. This view of the terrain was taken by the *Apollo 11* lunar module in 1969. One of the primary purposes for exploring the Moon was to bring back samples of rock to study them and to discover their origins. The Moon is made up of similar but not identical material to that found on the Earth. There is less iron on the Moon, but the major minerals are silicates as they are on the Earth (p. 43) – though they are slightly different in composition. This has led to the latest theory about the Moon's origins, sometimes known as the "Big Splash" theory. The impact of a large planet tore away parts of the Earth and parts of the large planet to create the unique geological makeup of the Moon.

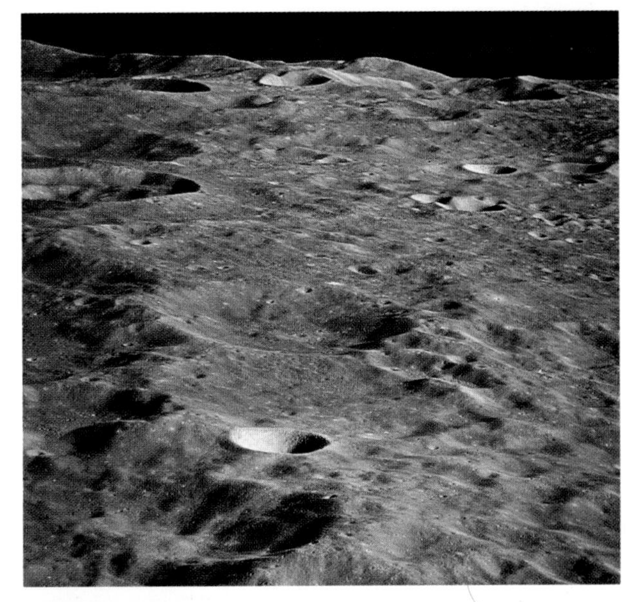

INVESTIGATING MOON ROCK
Rocks from the Moon have been investigated by geologists in the same way as they study Earth rocks. The rocks are ground down to thin slices and then looked at under a powerful microscope. The minerals, chiefly feldspar and olivine, which are abundant on the Earth, are unweathered. This is exceptional for geologists because there are no Earth rocks that are totally unweathered.

Cross polarized light in the microscope gives colors

Watery clearness shows no weathering

FACTS ABOUT THE MOON

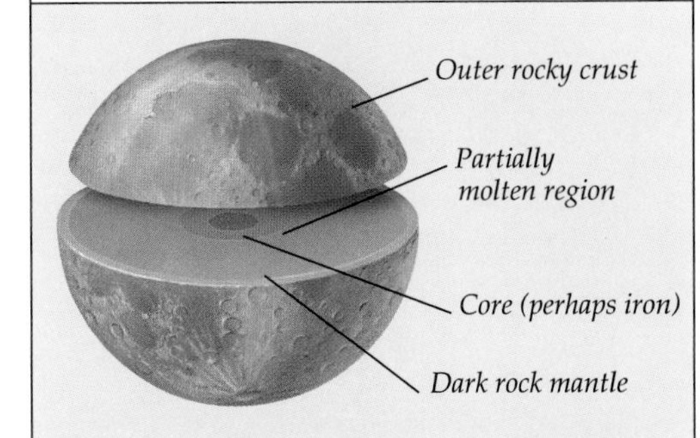

- **Interval between two New Moons** 29 days 12 hr 44 min
- **Temperature at surface** –230°F to +220°F
- **Rotational period** 27.3 Earth days
- **Mean distance from the Earth** 239,000 miles/384,000 km
- **Volume** (Earth = 1) 0.02
- **Mass** (Earth = 1) 0.012
- **Density** (water = 1) 3.34
- **Equatorial diameter** 2,160 miles/3,480 km

A MOON GLOBE
Selenography is the study of the surface features of the Moon and this selenograph, created by the artist John Russell in 1797, is a Moon globe. Only a little more than half of the globe is filled with images because at that time the features on the far side of the Moon were unknown. Not until the Russians received the earliest transmissions from the *Luna 3* probe in October 1959 was possible to see images of what was on the Moon's far side.

The Earth

THE EARTH IS THE ONLY PLANET in the Solar System that seems capable of supporting life. Its unique combination of liquid water, a rich oxygen- and nitrogen-based atmosphere, and dynamic weather patterns provide the basic elements for a diverse distribution of plant and animal life. Over millions of years, landforms and oceans have been constantly changing, mountains have been raised up and eroded away, and continental plates have drifted across the Earth. Some scientists see this harmonious balance being threatened by overpopulation by human beings. The stripping of rain forests and burning fossil fuels mean that carbon dioxide is building up in the atmosphere faster than plant life can recycle it back into oxygen. Since carbon dioxide traps the heat of the Sun beneath the Earth's atmosphere, but stops it from getting out again ("greenhouse effect," p. 47), the Earth's temperature may rise.

THE EARTH AND THE MOON
The English astronomer James Bradley (1693-1762) noted that many stars appear to have irregularities in their paths. He deduced that this is due to the effect of observing from an Earth that wobbles on its axis, caused by the gravitational pull of the Moon (p. 41).

CONSTANT GEOGRAPHICAL CHANGE
The Earth's crust is made up of a number of plates that are constantly moving because of currents that rise and fall from the molten iron core at the center of the Earth. Where the plates collide, they can lift the rocky landscape upward to create mountain ranges that are then eroded into craggy shapes like the Andes in Patagonia. The tensions caused by these movements sometimes result in earthquakes and volcanic activity.

Sahara desert

Water covers two-thirds of the Earth's surface

Cloud layers

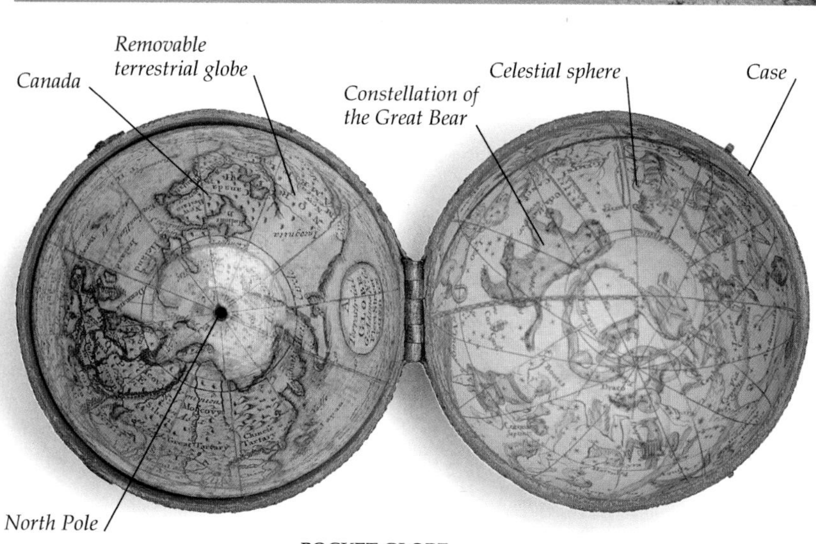

POCKET GLOBE
A globe is a convenient tool for recording specific features of the Earth's surface. This 19th-century pocket globe summarizes the face of the world from the geopolitical perspective, where the continents are divided into nations and spheres of influence. On the inside of the case is a map of the celestial sphere (pp. 12-13), with all the constellations marked out.

Collenia

FOSSILS IN ROCK

Dead creatures are buried in the sediment that is eroded from mountains, and they become fossils. This rock contains the remains of tiny algae that were one of the earliest life forms.

EARLY LIFE ON THE EARTH

The first life on the Earth was primitive plants that took carbon dioxide from the air and released oxygen during photosynthesis. Animals evolved when there was enough oxygen in the atmosphere to sustain them. Knowledge about evolving life forms comes in the form of fossils in the rocks (left). However, life forms survive only if environmental conditions on the Earth are suitable for them. The dinosaurs, for example, though perfectly adapted to their age became extinct about 65 million years ago.

Plant-eating *Edmontosaurus*

HUMAN DAMAGE

Many scientists wonder if humans, like the dinosaurs, might also become extinct. The dinosaur seems to have been a passive victim of the changing Earth, while humans are playing a key role in the destruction of their environment. By the year 2000 there will be more than 6 billion people on the Earth – all producing waste and pollution. In addition to global warming that may be occurring due to the greenhouse effect, chemicals are being released that deplete the ozone layer – a layer in the atmosphere that keeps out dangerous ultraviolet radiation.

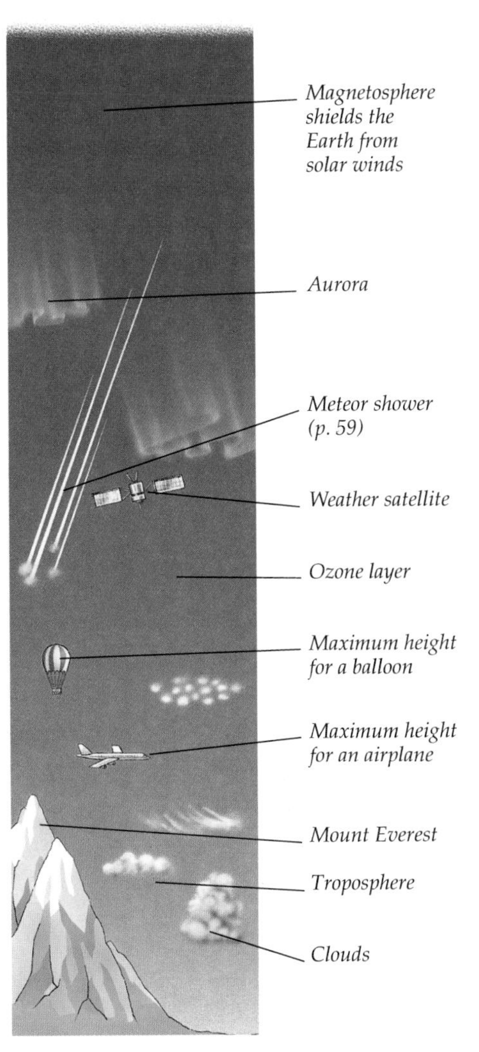

LIFE-GIVING ATMOSPHERE

Our atmosphere extends out for about 600 miles (1,000 km). It sustains life and protects us from the harmful effects of solar radiation. It has several layers, but the life-sustaining layer is the troposphere, up to 6 miles (10 km) above the Earth's surface.

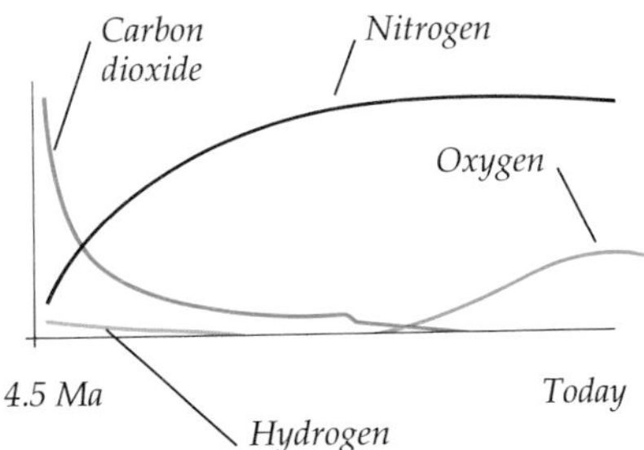

EVOLUTION OF THE ATMOSPHERE

Since the Earth was formed, the chemical make-up of the atmosphere has evolved. Carbon dioxide decreased significantly between 4.5 and 3 billion years ago (Ma). There was a comparable rise in nitrogen. The levels of oxygen began to rise at the same time, due to photosynthesis of primitive plants which used up CO_2 and released oxygen.

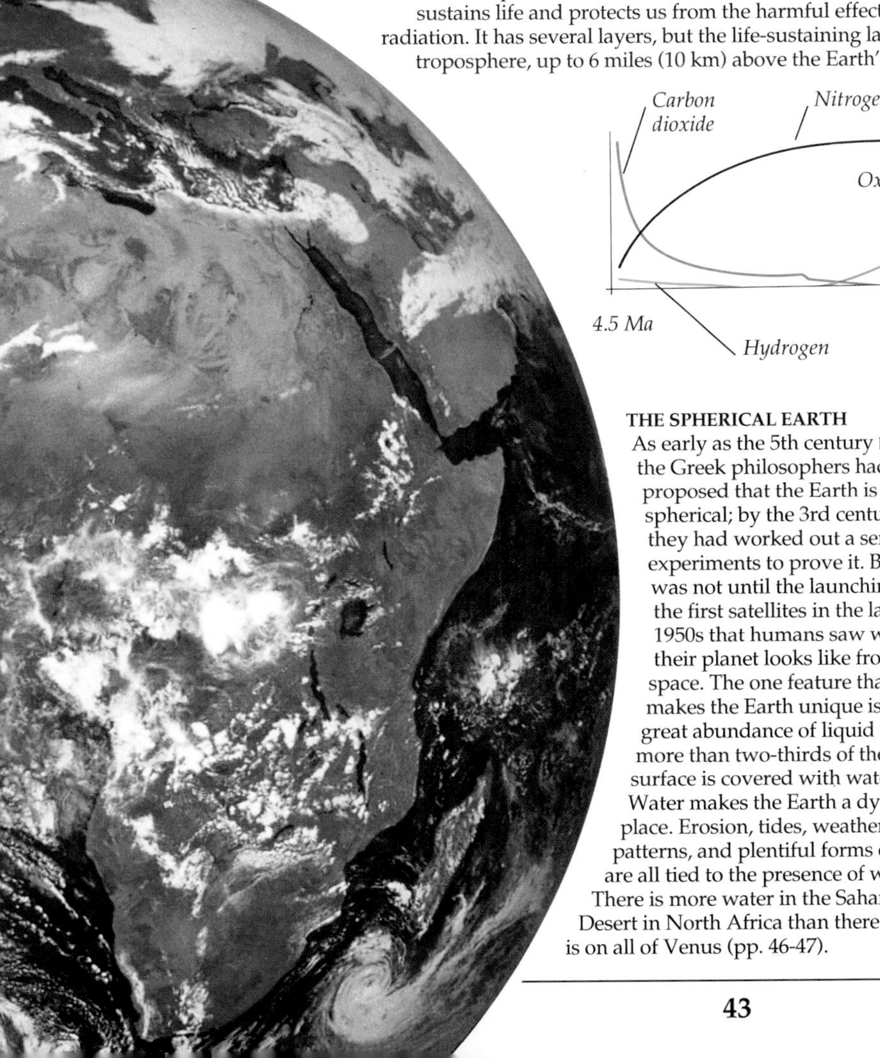

THE SPHERICAL EARTH

As early as the 5th century BC the Greek philosophers had proposed that the Earth is spherical; by the 3rd century BC, they had worked out a series of experiments to prove it. But it was not until the launching of the first satellites in the late 1950s that humans saw what their planet looks like from space. The one feature that makes the Earth unique is the great abundance of liquid water; more than two-thirds of the surface is covered with water. Water makes the Earth a dynamic place. Erosion, tides, weather patterns, and plentiful forms of life are all tied to the presence of water. There is more water in the Sahara Desert in North Africa than there is on all of Venus (pp. 46-47).

FACTS ABOUT THE EARTH

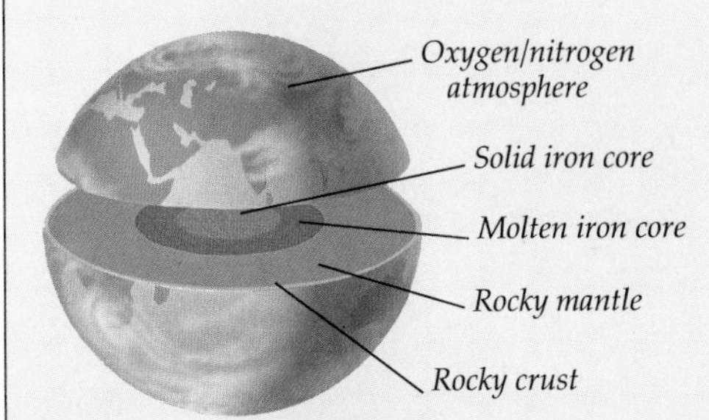

- **Sidereal period** 365.26 days
- **Temperature** –94°F to +130°F
- **Rotational period** 23 hr 56 min
- **Mean distance from the Sun** 93 million miles/ 150 million km
- **Volume** 1 • **Mass** 1
- **Density** (water = 1) 5.52
- **Equatorial diameter** 7,930 miles/12,760 km
- **Number of satellites** 1

Mercury

THE PLANET MERCURY IS NAMED after the Greco-Roman messenger of the gods, because it circles the Sun faster than the other planets, completing its circuit in 88 Earth days. Because it travels so close to the Sun, Mercury is often difficult to observe. Even though its reflected light makes it one of the brightest objects in the night sky, Mercury is never far enough from the Sun to be able to shine out brightly. It is only visible as a "morning" or "evening" star, hugging the horizon just before or after the Sun rises or sets. Like Venus, Mercury also has phases (p. 20). Being so close to the Sun, temperatures during the day on Mercury are hot enough to melt many metals. At night they drop to –279°F (–180°C), making the temperature range the greatest of all the planets. The gravitational pull of the Sun has "stolen" any atmosphere that Mercury had to protect itself against these extremes.

EARLY MERCURY MAP
Although many astronomers have tried to record the elusive face of Mercury, the most prolific observer was the French astronomer, Eugène Antoniadi (1870-1944). His maps, drawn between 1924 and 1929, show a number of huge valleys and deserts. Closeup views by the *Mariner 10* space probe uncovered an altogether different picture (below).

CRATERED TERRAIN
The surface of Mercury closely resembles our crater-covered Moon (p. 40). Mercury's craters were also formed by the impact of meteorites, and the lack of atmosphere has kept the landscape unchanged. Around the edges of the craters, a series of concentric ridges record how the surface was pushed outward by the force of the impact.

SEISMIC WAVES
Some of Mercury's hills and mountains were created by the impact of a huge meteorite (p. 59). The impact created a crater, known as Caloris Basin, where the meteorite struck the surface and sent out seismic, or shock, waves through the semi-molten core of the planet. These waves traveled through Mercury to the other side, where the crust buckled and mountain ranges were thrown up.

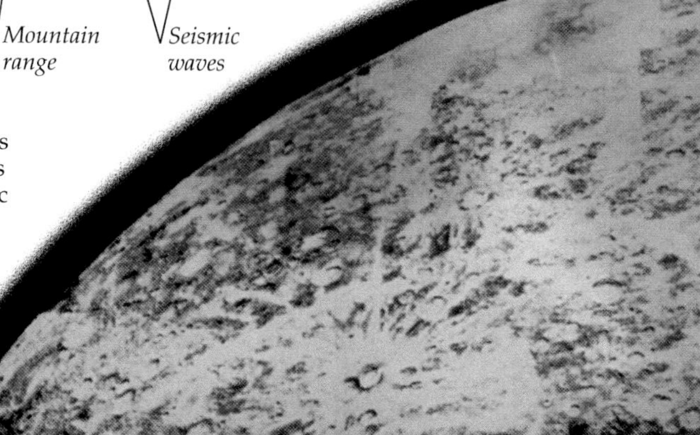

SURFACE OF MERCURY
This image is a mosaic of photographs taken during *Mariner 10*'s journey past Mercury in 1974. Mercury seems to have shrunk a bit after it was formed. This has caused a series of winding ridges, called scarps, that are unique to the planet. The entire surface is heavily cratered. The space probe *Mariner* also discovered that Mercury had a magnetic field about 1 percent the strength of the Earth's magnetic field.

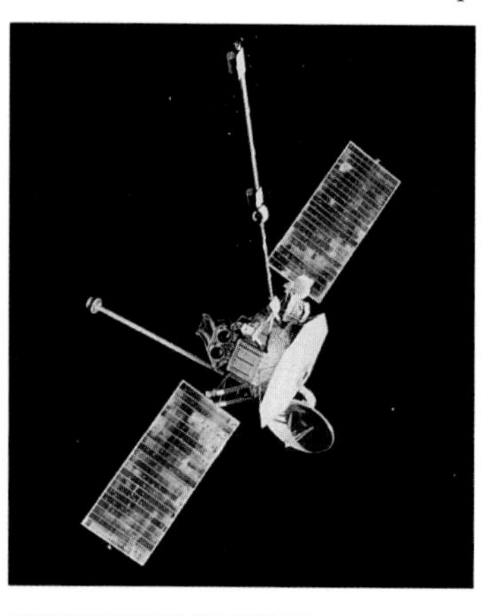

OBSERVING MERCURY
In 1973 the *Mariner 10* space probe was sent into orbit around the Sun where it behaved rather like a mini-planet. It sent back images of Mercury on three separate encounters before its cameras failed.

Measuring planets

Whereas we can weigh and measure objects on the Earth, we have to figure out the space a planet occupies (volume), how much matter it contains (mass), and its density by looking at its behavior, by analyzing its gravitational pull on nearby objects, and by using data gained by space probes (pp. 34-35). Density is the mass for every unit of volume of an object (mass divided by the volume).

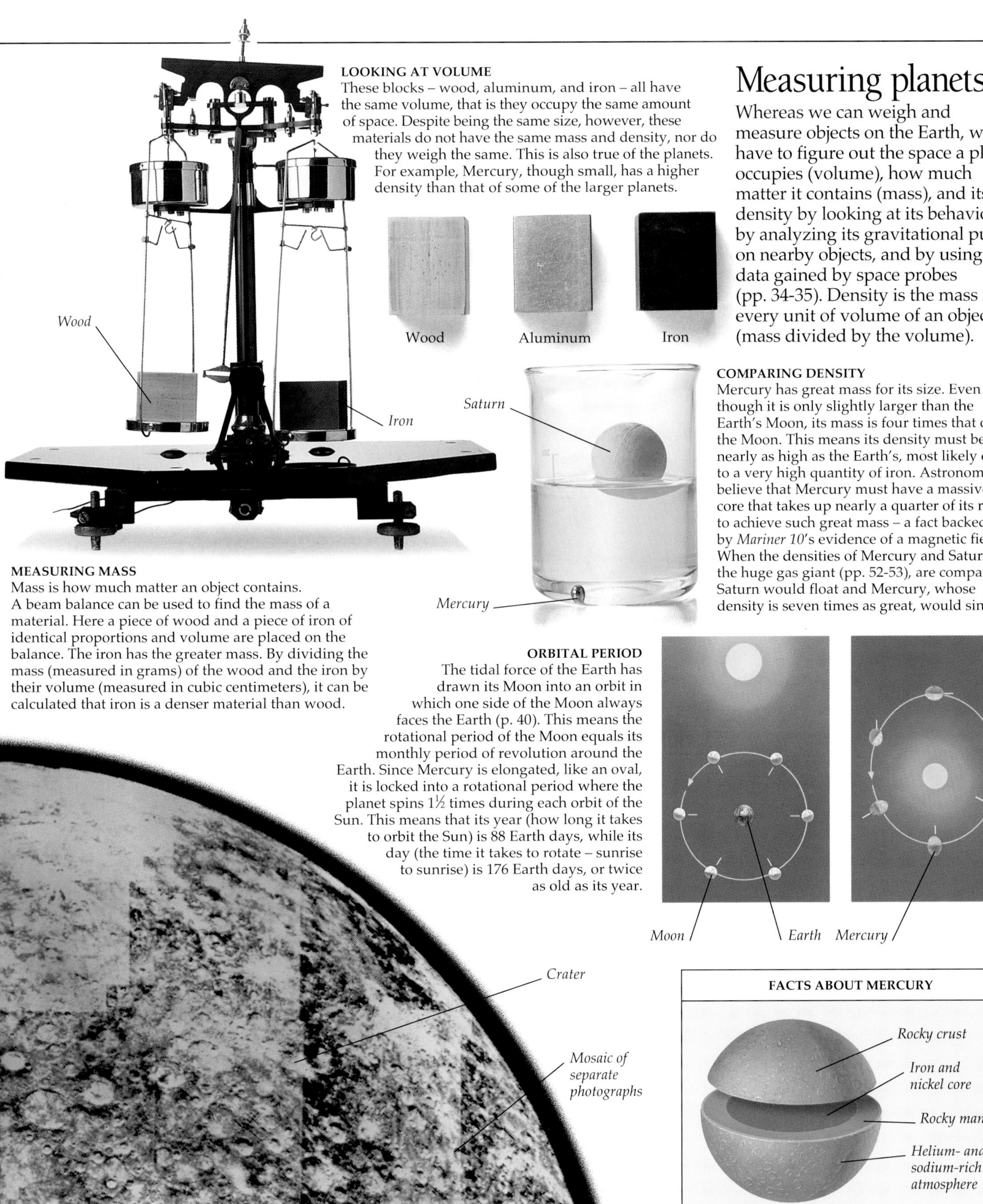

LOOKING AT VOLUME
These blocks – wood, aluminum, and iron – all have the same volume, that is they occupy the same amount of space. Despite being the same size, however, these materials do not have the same mass and density, nor do they weigh the same. This is also true of the planets. For example, Mercury, though small, has a higher density than that of some of the larger planets.

MEASURING MASS
Mass is how much matter an object contains. A beam balance can be used to find the mass of a material. Here a piece of wood and a piece of iron of identical proportions and volume are placed on the balance. The iron has the greater mass. By dividing the mass (measured in grams) of the wood and the iron by their volume (measured in cubic centimeters), it can be calculated that iron is a denser material than wood.

COMPARING DENSITY
Mercury has great mass for its size. Even though it is only slightly larger than the Earth's Moon, its mass is four times that of the Moon. This means its density must be nearly as high as the Earth's, most likely due to a very high quantity of iron. Astronomers believe that Mercury must have a massive iron core that takes up nearly a quarter of its radius to achieve such great mass – a fact backed up by *Mariner 10*'s evidence of a magnetic field. When the densities of Mercury and Saturn, the huge gas giant (pp. 52-53), are compared, Saturn would float and Mercury, whose density is seven times as great, would sink.

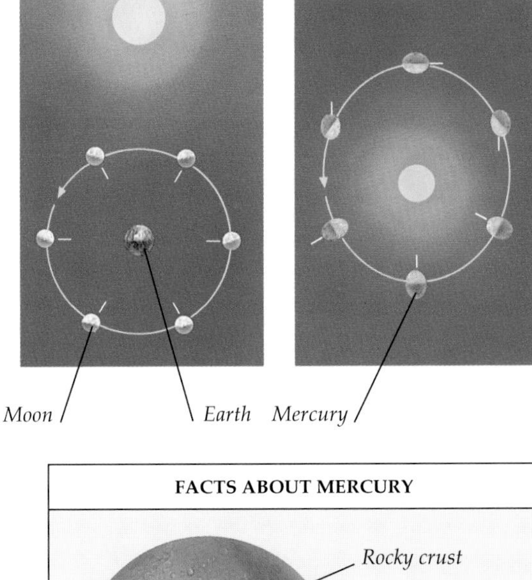

ORBITAL PERIOD
The tidal force of the Earth has drawn its Moon into an orbit in which one side of the Moon always faces the Earth (p. 40). This means the rotational period of the Moon equals its monthly period of revolution around the Earth. Since Mercury is elongated, like an oval, it is locked into a rotational period where the planet spins 1½ times during each orbit of the Sun. This means that its year (how long it takes to orbit the Sun) is 88 Earth days, while its day (the time it takes to rotate – sunrise to sunrise) is 176 Earth days, or twice as old as its year.

FACTS ABOUT MERCURY

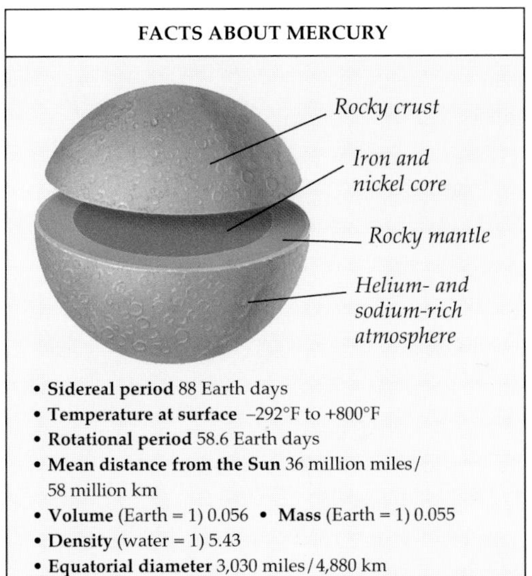

- **Sidereal period** 88 Earth days
- **Temperature at surface** –292°F to +800°F
- **Rotational period** 58.6 Earth days
- **Mean distance from the Sun** 36 million miles/ 58 million km
- **Volume** (Earth = 1) 0.056 • **Mass** (Earth = 1) 0.055
- **Density** (water = 1) 5.43
- **Equatorial diameter** 3,030 miles/4,880 km
- **Number of satellites** 0

Venus

PEOPLE OFTEN MISTAKE VENUS for a star. After the Moon, it is the brightest object in our night sky. Because it is so close in size to the Earth, until the 20th century astronomers assumed that it might be in some ways like the Earth. The probes sent to investigate have shown that this is not so. The dense cloudy atmosphere of Venus hides its surface from even the most powerful telescope. Only radar can penetrate to map the planet's features. Until it became possible to determine the surface features – largely flat, volcanic plains – scientists could not tell how long the Venusian day was. The atmosphere is deadly. It is made up of a mixture of carbon dioxide and sulfuric acid, which causes an extreme "greenhouse effect" in which heat is trapped by the lethal atmosphere. The ancients, however, saw only a beautifully bright planet and so they named it after their goddess of love. Nearly all the features mapped on the surface of Venus have been named after women, such as Pavlova, Sappho, and Phoebe.

VENUS IN THE NIGHT SKY
This photograph was taken from the Earth. It shows the crescent Moon with Venus in the upper left of the sky. Shining like a lantern at twilight, Venus looks so attractive that some astronomers were inspired to believe it must be a beautiful planet.

CALCULATING DISTANCES
One way to calculate the distance of the Earth from the Sun is for a number of observers all around the world to measure the transit of a planet (the passage of the planet as it crosses the disc of the Sun and appears in silhouette). The British explorer Captain James Cook led one of the many expeditions in 1769 to observe the Transit of Venus from Tahiti. Calculations made from these observations also enabled astronomers to work out the relative measurements of the entire Solar System.

FACTS ABOUT VENUS

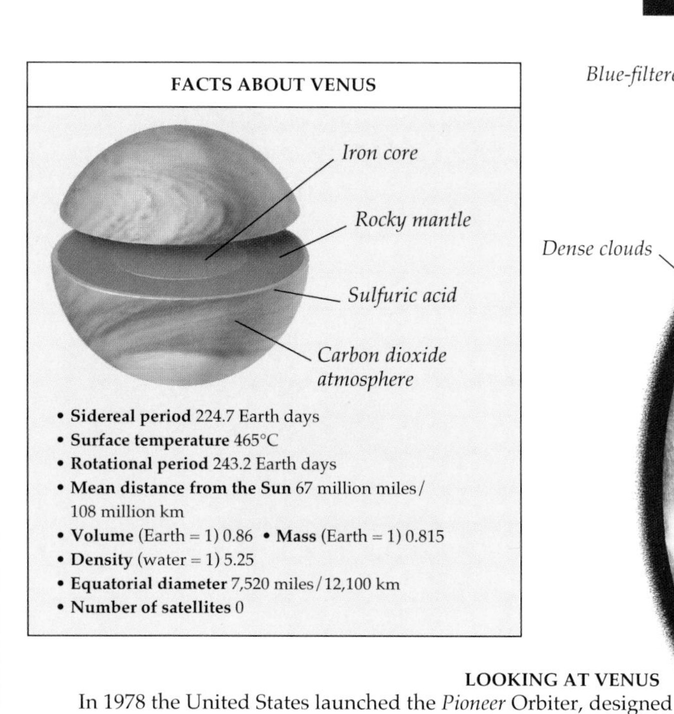

- **Sidereal period** 224.7 Earth days
- **Surface temperature** 465°C
- **Rotational period** 243.2 Earth days
- **Mean distance from the Sun** 67 million miles/108 million km
- **Volume** (Earth = 1) 0.86 • **Mass** (Earth = 1) 0.815
- **Density** (water = 1) 5.25
- **Equatorial diameter** 7,520 miles/12,100 km
- **Number of satellites** 0

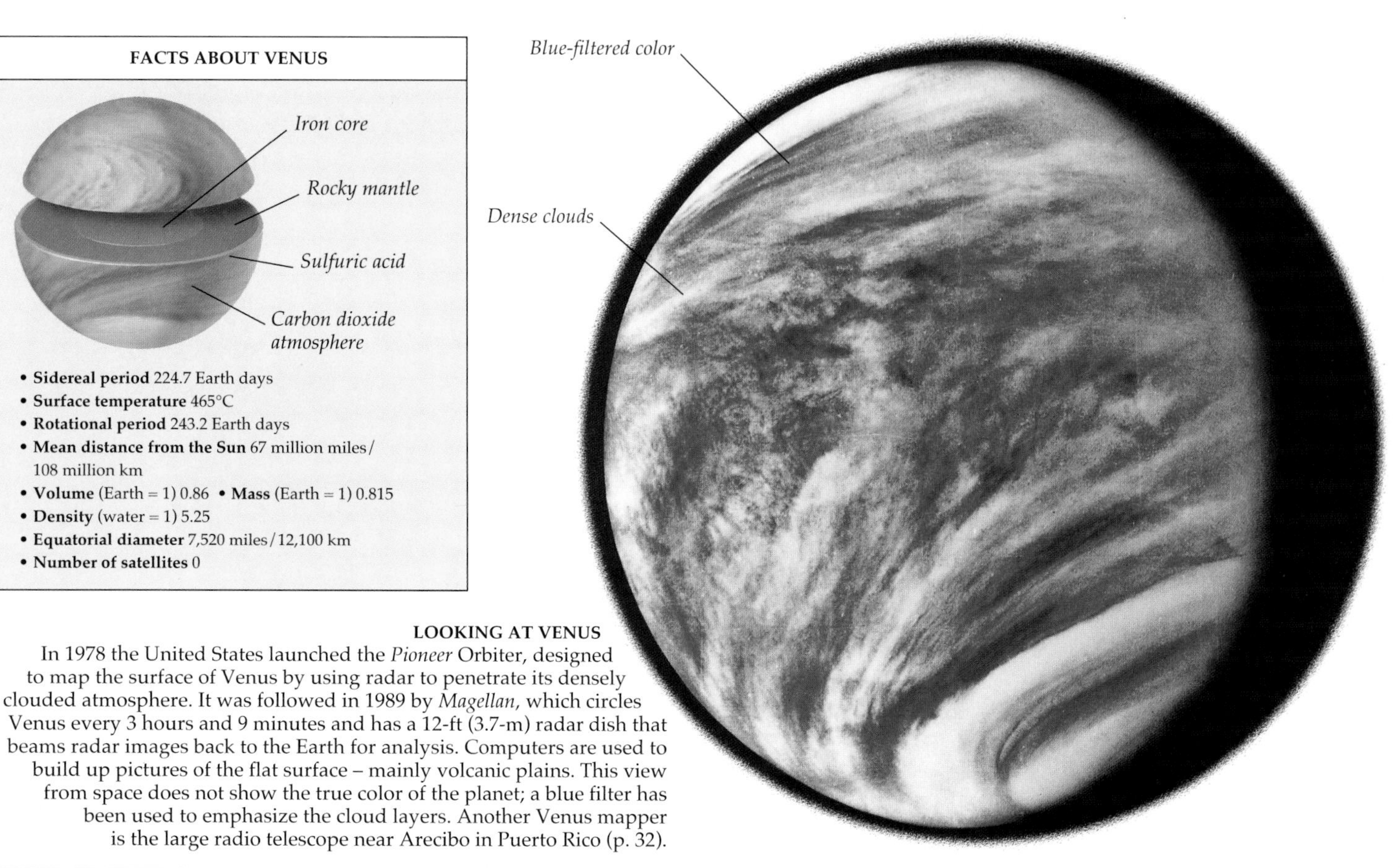

LOOKING AT VENUS
In 1978 the United States launched the *Pioneer* Orbiter, designed to map the surface of Venus by using radar to penetrate its densely clouded atmosphere. It was followed in 1989 by *Magellan*, which circles Venus every 3 hours and 9 minutes and has a 12-ft (3.7-m) radar dish that beams radar images back to the Earth for analysis. Computers are used to build up pictures of the flat surface – mainly volcanic plains. This view from space does not show the true color of the planet; a blue filter has been used to emphasize the cloud layers. Another Venus mapper is the large radio telescope near Arecibo in Puerto Rico (p. 32).

PUZZLING SURFACE
Even with the best telescope, Venus looks almost blank. This led the Russian astronomer, Mikhail Lomonosov (p. 28) to propose that the Venusian surface is densely covered with cloud. As recently as 1955 the British astronomer Fred Hoyle (1915-) argued that the clouds are actually drops of oil and that the oceans on Venus are oceans of oil. Radar has shown that Venus's surface is actually covered with live volcanoes.

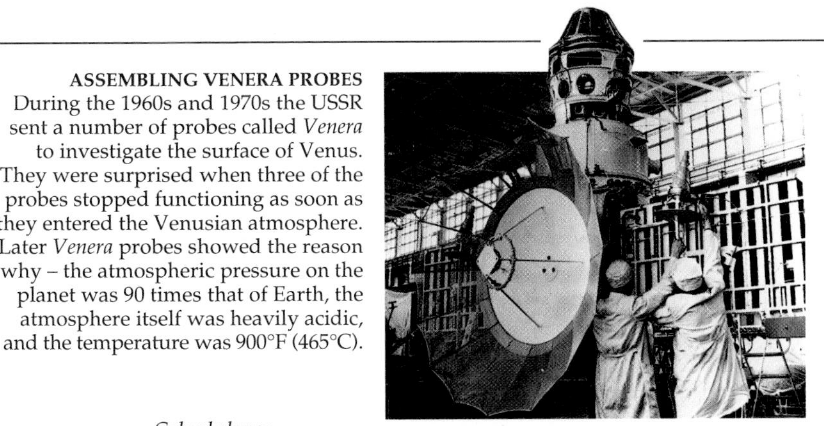

ASSEMBLING VENERA PROBES
During the 1960s and 1970s the USSR sent a number of probes called *Venera* to investigate the surface of Venus. They were surprised when three of the probes stopped functioning as soon as they entered the Venusian atmosphere. Later *Venera* probes showed the reason why – the atmospheric pressure on the planet was 90 times that of Earth, the atmosphere itself was heavily acidic, and the temperature was 900°F (465°C).

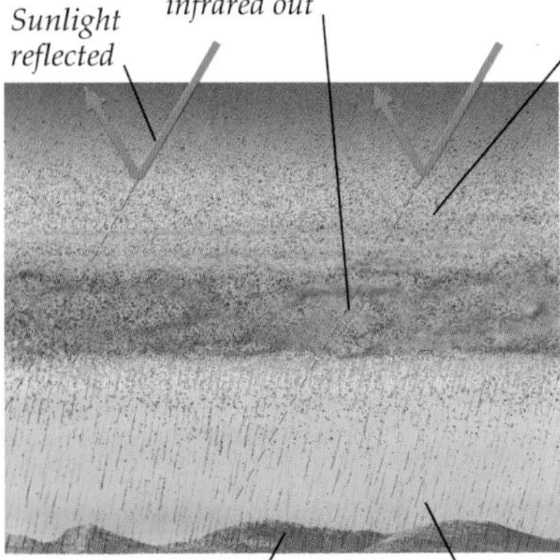

GREENHOUSE EFFECT
The great amount of carbon dioxide in Venus's atmosphere means that, while sunlight can penetrate, heat cannot escape. This has led to a runaway "greenhouse effect." Temperatures on the surface easily reach 900°F (465°C), even though the thick cloud layers keep out as much as 80 percent of the Sun's rays.

Color balance

Feet of probe

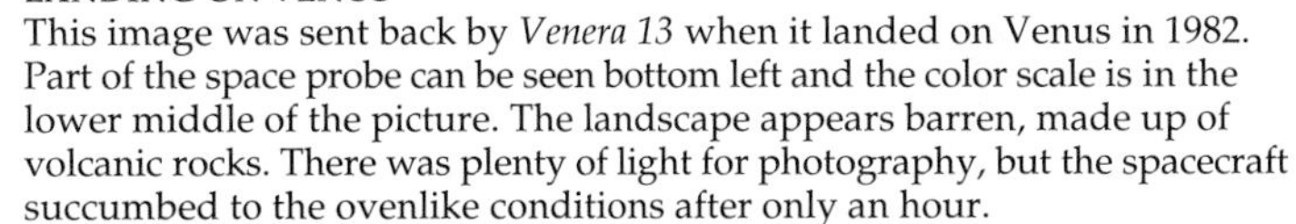

LANDING ON VENUS
This image was sent back by *Venera 13* when it landed on Venus in 1982. Part of the space probe can be seen bottom left and the color scale is in the lower middle of the picture. The landscape appears barren, made up of volcanic rocks. There was plenty of light for photography, but the spacecraft succumbed to the ovenlike conditions after only an hour.

Sif Mons volcano

Gula Mons volcano

Lava flows

THREE-DIMENSIONAL VIEW
This image of the Western Eistla Region, sent by *Magellan*, shows the volcanic lava flows (see here as the bright features) that cover the landscape and blanket the original Venusian features. Most of the landscape is covered by shallow craters. The simulated colors are based on those recorded by the Soviet *Venera* probes.

Mars

MARS APPEARS SOMEWHAT REDDISH IN THE NIGHT SKY. The Babylonians, Greeks, and Romans all named it after their gods of war. In reality, Mars is a small planet – only half the size of the Earth – but there are similarities. Mars, like the Earth, has a $24\frac{1}{2}$-hour day, polar caps, and an atmosphere. Not surprisingly, Mars has always been the most popular candidate as a site for possible extraterrestrial life. But, despite the similarities with the Earth, there now seems little chance that there is life on Mars. The atmosphere is too thin to support life and allows deadly ultraviolet rays to reach the surface. Mars is also farther from the Sun than the Earth, making it a lot colder.

MARTIAN MARKINGS
In 1659 the Dutch scientist Christiaan Huygens (1629-1695) drew the first map of Mars, showing a V-shaped mark on the surface that reappeared in the same place every 24 hours. This was Syrtis Major. He concluded, correctly, that its regular appearance indicated the length of the Martian day. The American astronomer, Percival Lowell (1855-1916) made a beautiful series of drawings of the Martian canals described by Schiaparelli (below). Closer inspections showed that these canals are caused by the effects of observing through his telescope with the lens closed down too far.

Arabia region

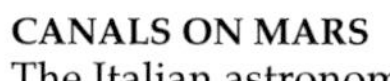

CANALS ON MARS
The Italian astronomer Giovanni Schiaparelli (1835-1910) made a close study of the surface of Mars. In 1877 he noticed a series of dark lines that seemed to form some sort of network. Schiaparelli called them *canali*, translated as "channels" or "canals." This optical illusion seems to be the origin of the myth that Mars is occupied by a sophisticated race of hydraulic engineers. It was Eugène Antoniadi (p. 44) who made the first accurate map of Mars.

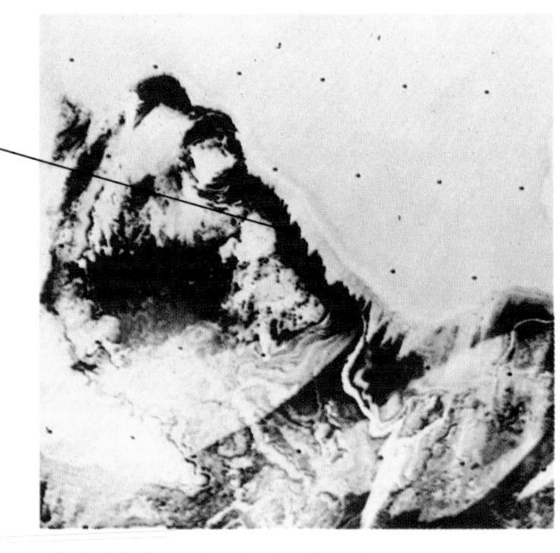

Ice cliff

Ice cap

AROUND THE PLANET MARS
The Martian atmosphere is much thinner than that of the Earth and is composed mostly of carbon dioxide. This is because Mars has a low gravitational attraction and the lighter gases have escaped from its atmosphere into space (p. 36). The *Viking 2* space probe recorded a noticeable amount of water vapour and ice cliffs (right), and photographs taken by the first orbiting space probe, *Mariner 9*, showed a series of winding valleys in the Chryse region that could have been dried-up river beds. Mars also has large volcanoes. One of them – Olympus Mons – is the largest in the Solar System. There are also deserts, canyons, and polar ice caps.

EVIDENCE OF POLAR ICE
What little water there is on Mars seems to be permanently locked in the frozen ice caps. Mars has white markings covering both of its poles. These seem to shrink and expand like our polar caps on the Earth. The Martian ice caps are largely composed of water ice and some frozen carbon dioxide. These ice cliffs on the northern polar cap are up to 2 miles (3.5 km) thick.

Computer-processed view of Mars from Viking orbiters (1980)

SURFACE OF MARS
The landscape on Mars is strewn with rubble – everything is covered with a deep layer of rust-colored iron oxide dust. The arm of the *Viking* lander can be seen scooping a sample of rock.

TESTING FOR LIFE
The two *Viking* probes in the 1970s carried out simple experiments on Martian soil. They found no signs of life.

Assembling the Viking lander

ASAPH HALL
The two Martian moons were discovered in 1877 by the American astronomer Asaph Hall (1829-1907). They are probably rogue asteroids caught in Mars's gravity.

DESERT LANDSCAPE
Mars resembles a desert. Winds whip up the dust into storms and the red dust becomes suspended in the atmosphere, giving the sky a reddish hue. The images of the landscape taken by the space probe can be colored on film later using a strip of paint chips on board that is photographed at the same time as the landscape (p. 47). In this way the known colors of the paint chips are used to help make up the colors on the film back in the laboratory.

MARTIAN MOONS
Mars has two small moons, Phobos (right) and Deimos, 17 and 10 miles (28 and 16 km) in diameter. Since the orbit of Deimos is only 14,580 miles (23,460 km) from the center of Mars, it will probably be pulled down to the surface with a crash in about 50 million years.

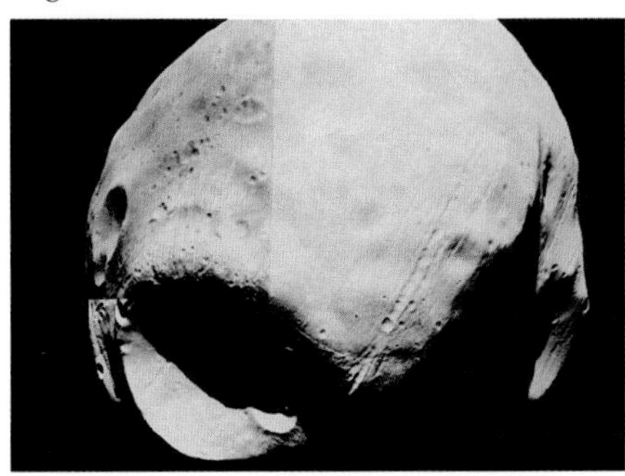

VALLES MARINERIS
The Valles Marineris are a network of canyons that run around nearly one-third of the equatorial region on Mars. There is evidence of landslides and erosion, with rock debris along its length. Some of the canyons are so self-contained that scientists believe there could never have been any water in them. Because Mars is not subjected to any plate movements (pp. 42-43), the landscape remains the same.

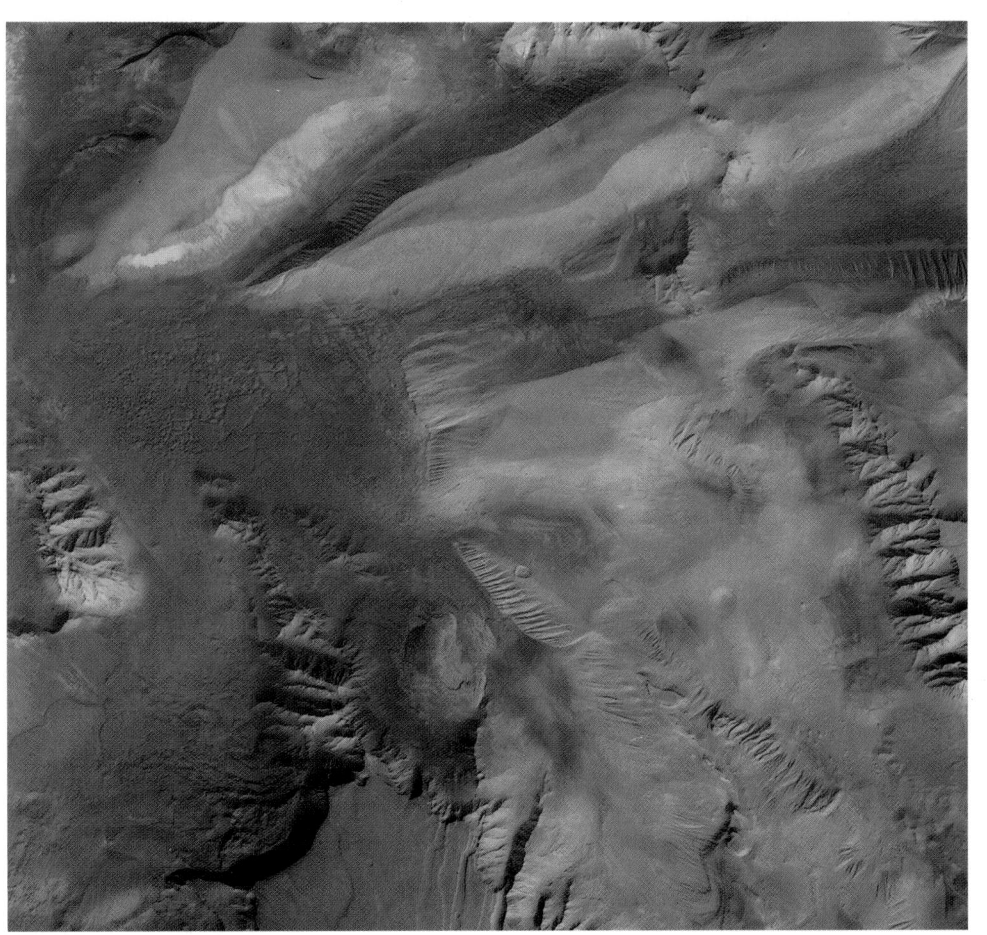

FACTS ABOUT MARS

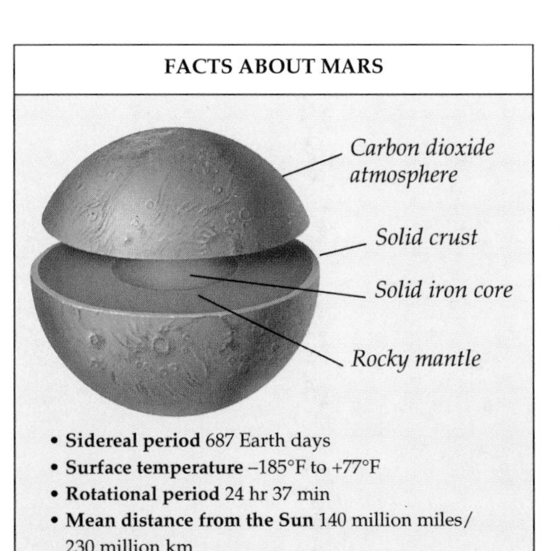

- **Sidereal period** 687 Earth days
- **Surface temperature** –185°F to +77°F
- **Rotational period** 24 hr 37 min
- **Mean distance from the Sun** 140 million miles/ 230 million km
- **Volume** (Earth=1) 0.15
- **Mass** (Earth=1) 0.11
- **Density** (water=1) 3.95
- **Equatorial diameter** 4,220 miles/6,790 km
- **Number of satellites** 2

Jupiter

THIS HUGE, BRIGHT PLANET is the largest world in our Solar System; four of its moons are the size of planets. It is different in structure from the solid inner planets. Most of Jupiter's bulk is gas – mainly hydrogen and helium. Below the level of the clouds, the pressure is so great that the hydrogen is compressed into its liquid form and, farther down, it turns into metallic hydrogen. It emits more heat radiation than it receives from the Sun, because it continues shrinking at a rate of a few millimeters a year. Had Jupiter been many times more massive, the gas at its core would have been hot and dense enough for nuclear fusion reactions to take place (p. 38), and Jupiter would have become a star. Even though Jupiter has remained a planet, space probes, which take several years to get there, must be protected from Jupiter's fierce radiation belts, which could destroy their sensitive equipment. Any probe sent into the turbulent atmosphere will eventually buckle like a tin can under the pressure.

JUPITER'S RINGS
The U.S. *Pioneer* missions were sent past Jupiter in the early 1970s, *Pioneer 10* sending back the first pictures. In 1977 the US sent two *Voyager* probes to explore Jupiter's cloud tops and five of its moons. *Voyager 1* uncovered a faint ring – like Saturn's rings (p. 53) – circling the planet. The thin yellow ring (approximately 18 miles/30 km thick) can be seen at the top of the photograph.

SEEING THE RED SPOT
In 1660 the English scientist Robert Hooke (1635-1702) reported seeing "a spot in the largest of the three belts of Jupiter." Gian Cassini (p. 28) saw the spot at the same time, but subsequent astronomers were unable to find it. The Great Red Spot was observed again in 1878 by the American astronomer Edward Barnard (1857-1923).

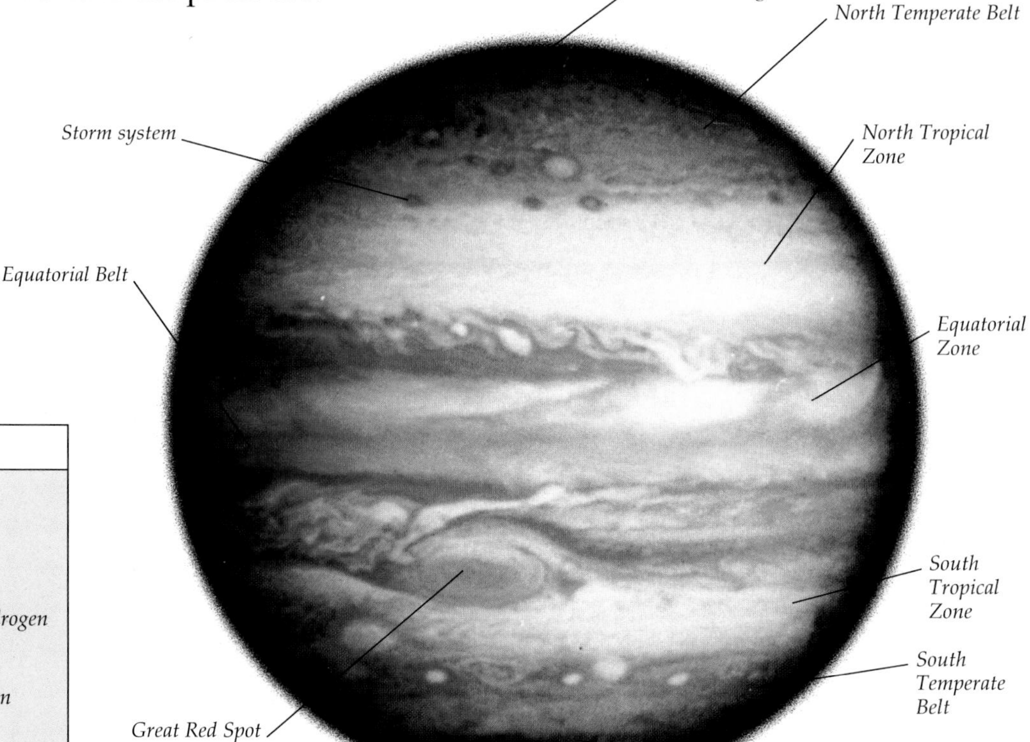

JUPITER'S CLOUDS
The cloud tops of Jupiter seem to be divided into a series of bands that are different colors. The light bands are called zones, and the dark bands belts. The North Tropical Zone (equivalent to our northern temperate zone) is the brightest, its whiteness indicating high-level ammonia clouds. The Equatorial Belt, surrounding Jupiter's equator, always seems in turmoil, with the atmosphere constantly whipped up by violent winds. Across the planet are a number of white or red ovals. These are huge cloud systems. The brown and orange bands indicate the presence of organic molecules, including ethane.

FACTS ABOUT JUPITER

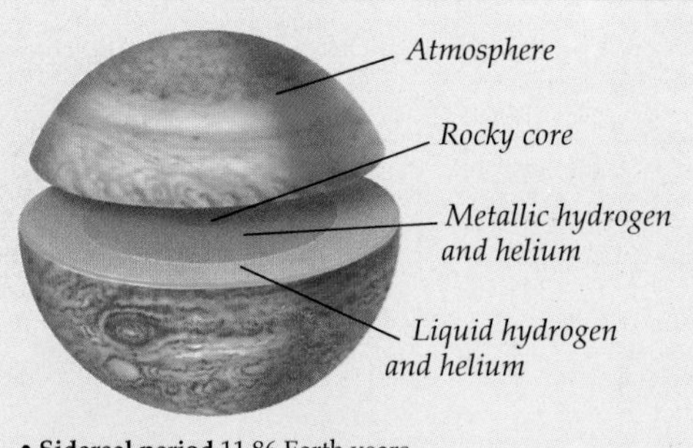

- **Sidereal period** 11.86 Earth years
- **Temperature at cloud tops** –240°F
- **Rotational period** 9 hr 55 min
- **Mean distance from the Sun** 483 million miles/ 778 million km
- **Volume** (Earth = 1) 1,319 • **Mass** (Earth = 1) 318
- **Density** (water = 1) 1.33
- **Equatorial diameter** 89,350 miles/142,980 km
- **Number of satellites** 16

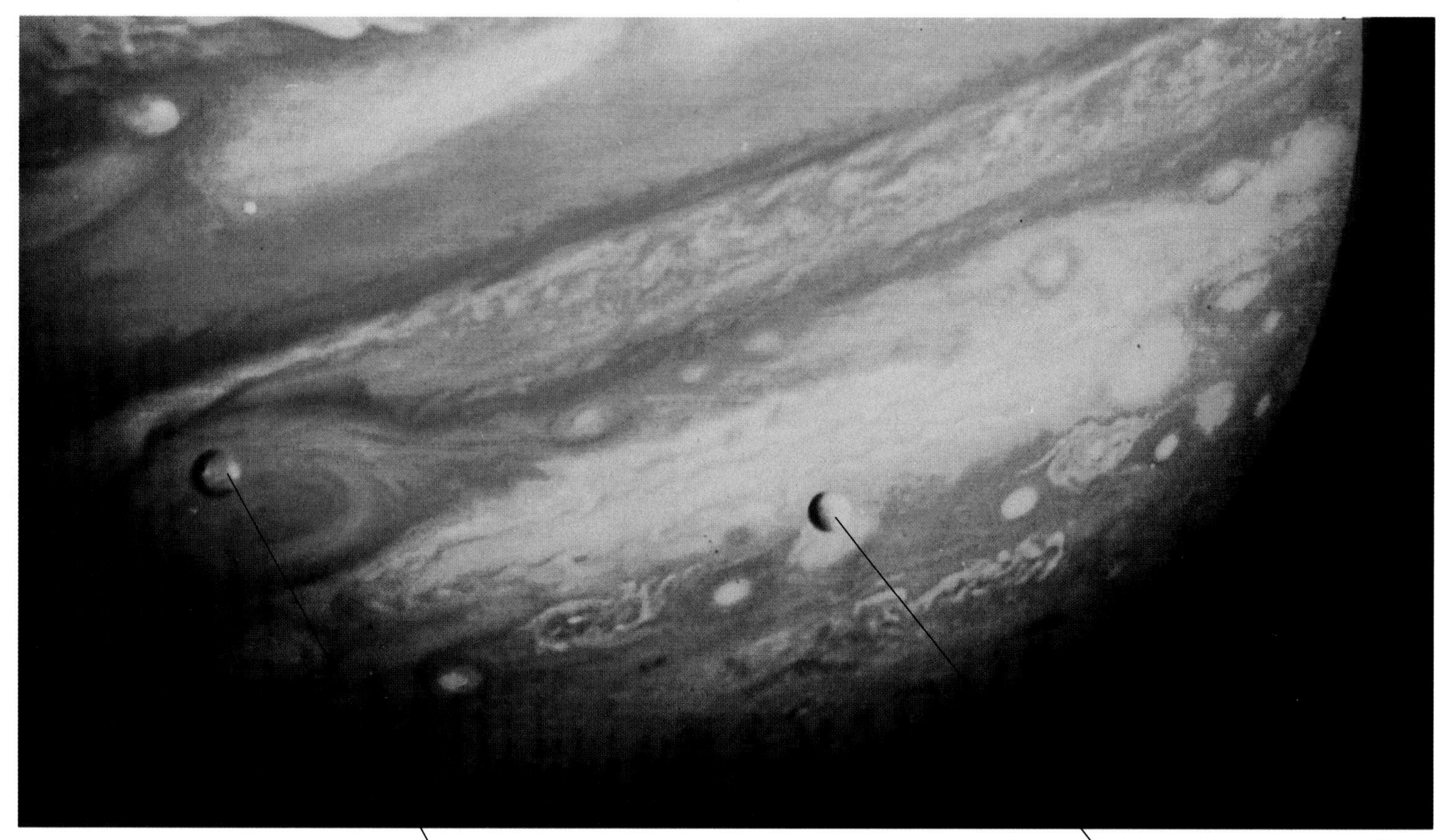

THE GREAT RED SPOT
Many spots are visible on Jupiter; most of them are short-lived except for the Great Red Spot (far left), a high-pressure storm system that was "rediscovered" in 1878, suggesting that the visibility of the Great Red Spot varies with time. The Great Red Spot stays at the same latitude; at present it is nearly 24,856 miles (40,000 km) across. Its red color is probably due to phosphorus, which usually lies close to the core of the planet, but is constantly being carried upward through the atmosphere by the turbulence of the storm system.

Jupiter's moons

In 1610 Galileo (p. 20) made the first systematic study of the four largest moons of Jupiter. Since they seemed to change their positions relative to the planet every night, he concluded, correctly, that these objects must be revolving around Jupiter. This insight provided more ammunition for the dismantling of the geocentric theory (p. 10), which placed the Earth at the center of the Universe. In 1892 another small moon was discovered circling close to the cloud tops of the planet. Since then, another 11 tiny moons have been discovered.

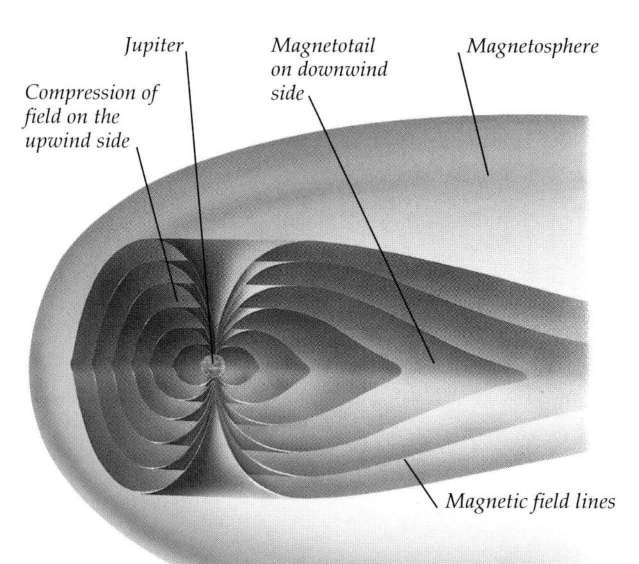

SPINNING JUPITER
Jupiter spins so quickly that its day is only 9 hours and 55 minutes long and its equator bulges outward. Another effect of the rapid rotation is that the spinning of Jupiter's metallic hydrogen core generates a huge magnetic field around the planet. This magnetosphere is buffeted and pushed back by the solar wind and its tail spreads out over a vast distance, away from the Sun.

CALLISTO
Callisto is the second largest of Jupiter's moons, and the most heavily cratered, not unlike our Moon, except that the craters are made of ice. The bright areas are the ice craters formed by impact of objects from space..

ERUPTION ON IO
Io is the Moon that is closest to Jupiter. It is one of the "Galilean" moons, named after Galileo, who discovered them. The others are Callisto, Europa, and Ganymede. The erupting plume of a massive volcano can be seen here on the horizon, throwing sulfuric materials 186 miles (300 km) out into space. The photograph was taken by *Voyager* from a distance of 310,700 miles (500,000 km) and has been specially colored using filters.

17TH-CENTURY VIEW
In 1675 the Bolognese director of the Paris Observatory, Gian Domenico Cassini (p. 50), discovered that Saturn's ring does not appear to be solid. It seemed that there are two rings, with a darkly hued gap separating the inner from the outer ring. His drawing, made in 1676, shows the gap, called Cassini's Division in his honor.

Saturn

THE GIANT PLANET SATURN, with its flat rings, is probably the most widely recognized astronomical image. For the classical world, Saturn was the most distant planet known. They named it after the original father of all the gods. Early astronomers noted its 29-year orbit and assumed that it moves sluggishly. Composed mostly of hydrogen, its atmosphere and structure are similar to Jupiter's, but its density is much lower. Saturn is so light that it could float on water (p. 45). Like Jupiter, Saturn rotates at great speed causing its equator to bulge outwards. Saturn also has an appreciable magnetic field. Winds in Saturn's upper atmosphere can travel at 1,800 km/h (1,100 mph), but only one major storm system has been discovered so far, Anne's Spot, named after *Voyager* scientist Anne Bunker.

SATURN AND THE RINGS
Saturn's rings appear solid from the Earth, and the *Voyager* probes showed that they are composed of different-sized chunks of ice. The high reflectivity of these ice particles helps explain Saturn's luminosity in the night sky. The rings themselves are only about 98 ft (30 m) thick, but the total width of the rings is more than 169,000 miles (272,000 km).

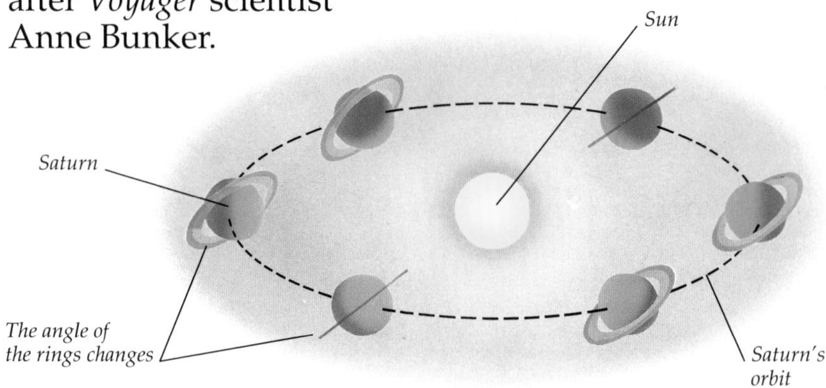

CHANGING VIEW
Saturn's axis is tilted. Because the rings lie around its equator, they incline as the planet tilts. This means that the rings change dramatically in appearance, depending on what time during Saturn's year they are being observed – Saturn's year is equal to 29.5 Earth years. The angle of the rings appears to change according to how Saturn and the Earth are placed in their respective orbits. When the rings are edge-on they look like a triple planet, as Galileo found in 1612.

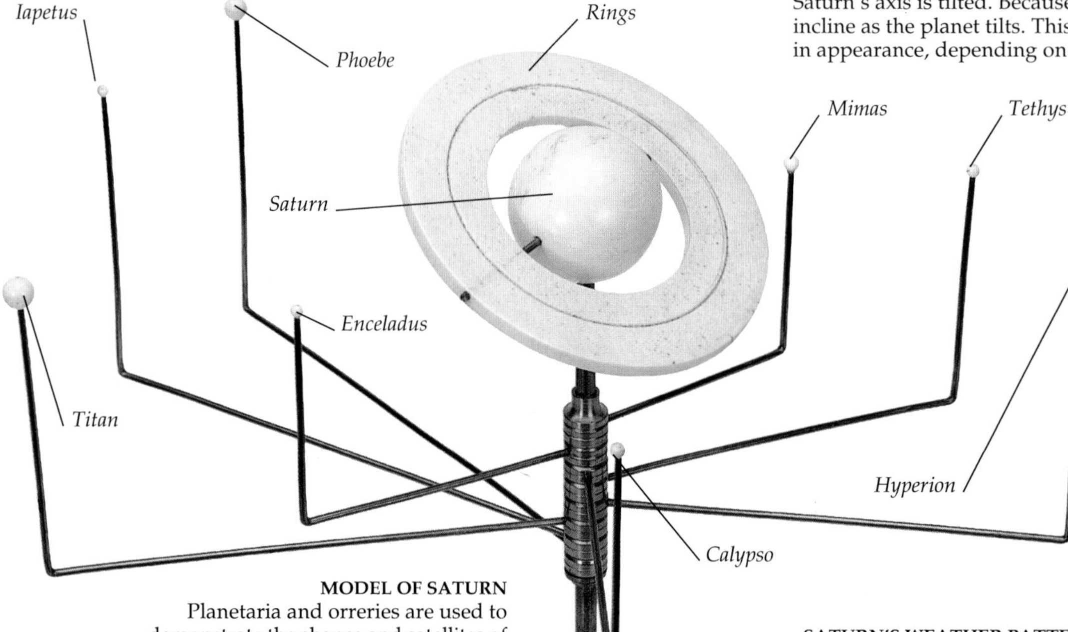

MODEL OF SATURN
Planetaria and orreries are used to demonstrate the shapes and satellites of the planets. This planetarium shows Saturn with the eight moons that were known in the 19th century. The model is flawed, however, because it is impossible to show the relative size of a planet and the orbits of its satellites.

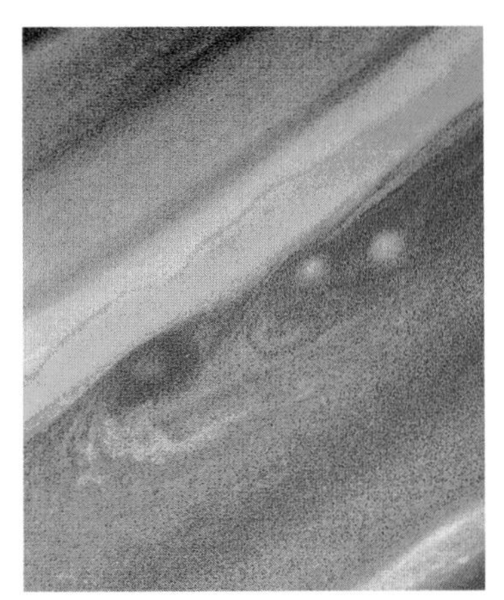

SATURN'S WEATHER PATTERNS
The weather patterns of Saturn's northern hemisphere were photographed from a range of 4 million miles (7 million km) by *Voyager 2* in 1981. Storm clouds and white spots are visible features of Saturn's weather.

SEEING THE RINGS
When Galileo first discovered Saturn's rings in 1610, he misinterpreted what he saw. He thought Saturn was a triple planet. It was not until 1655 that the rings were successfully identified and described by the Dutch scientist and astronomer Christiaan Huygens (1629-1695), using a powerful telescope that he built himself.

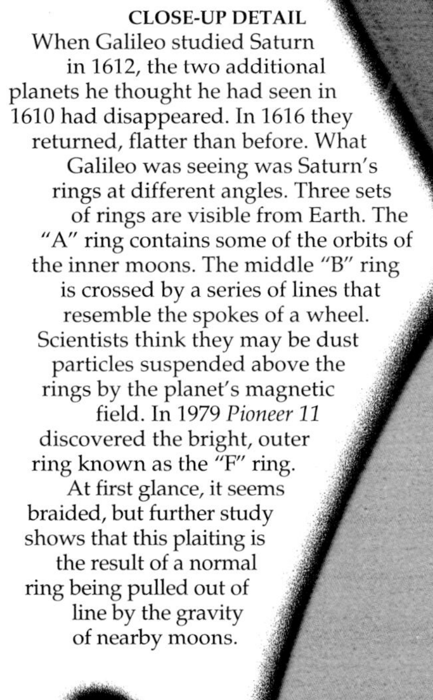

CLOSE-UP DETAIL
When Galileo studied Saturn in 1612, the two additional planets he thought he had seen in 1610 had disappeared. In 1616 they returned, flatter than before. What Galileo was seeing was Saturn's rings at different angles. Three sets of rings are visible from Earth. The "A" ring contains some of the orbits of the inner moons. The middle "B" ring is crossed by a series of lines that resemble the spokes of a wheel. Scientists think they may be dust particles suspended above the rings by the planet's magnetic field. In 1979 *Pioneer 11* discovered the bright, outer ring known as the "F" ring. At first glance, it seems braided, but further study shows that this plaiting is the result of a normal ring being pulled out of line by the gravity of nearby moons.

SATURN'S MOONS
All of Saturn's moons except the largest, Titan, are made of ice. Enceladus is a relatively small moon, and it shows evidence of recent change. Some parts of the moon are heavily cratered, while others are smooth, suggesting internal melting and subsequent surface melting. Enceladus has a diameter of 310 miles (500 km). This photograph was taken at a distance of 74,000 miles (119,000 km); features as small as 2 miles (3 km) can be identified.

FACTS ABOUT SATURN

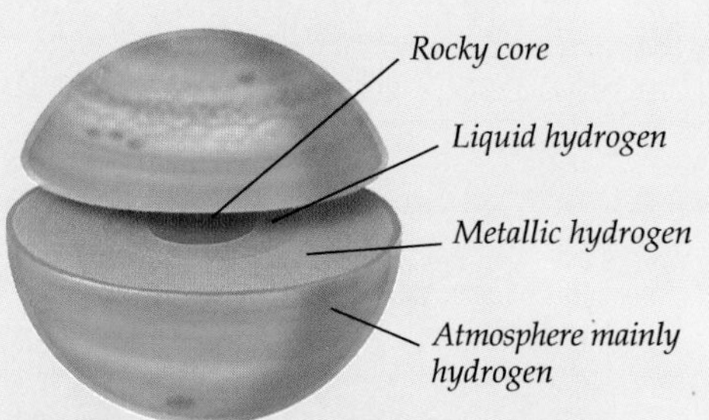

- **Sidereal period** 29.5 Earth years
- **Temperature at cloud tops** –292°F
- **Rotational period** 10 hr 40 min
- **Mean distance from the Sun** 886 million miles/ 1,430 million km
- **Volume** (Earth = 1) 744 • **Mass** (Earth = 1) 95.18
- **Density** (water = 1) 0.69
- **Equatorial diameter** 74,900 miles/120,540 km
- **Number of satellites** 18

Uranus

URANUS WAS THE FIRST PLANET to be discovered since the use of the telescope. Astronomers believed there must be another planet beyond Saturn because of Bode's Law – a mathematical formula that predicted roughly where planets should lie. Uranus was discovered when William Herschel, observing from Bath, England, set about re-measuring all the major stars with his 7-ft (2.1-m) reflector telescope (p. 24). In 1781 he noticed an unusually bright object in the zodiacal constellation of Gemini. At first he assumed it was a nebula (pp. 60-61) and then a comet (pp. 58-59), but it moved in a peculiar way. The name of Uranus was suggested by the German astronomer, Johann Bode (of Bode's Law), who proposed that the planet be named after the father of Saturn, in line with established classical traditions.

WILLIAM HERSCHEL (1738-1822)
William Herschel was so impressed by a treatise on optics, which described the construction of telescopes, that he wanted to buy his own telescope. He found them too expensive, so in 1773 he decided to start building his own. From that moment on, astronomy became Herschel's passion.

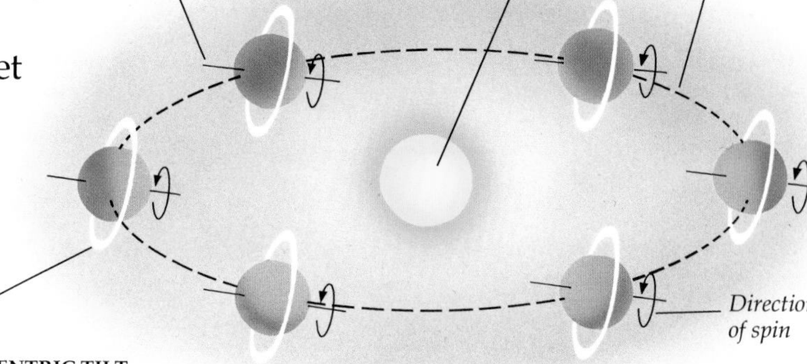

ECCENTRIC TILT
Uranus spins on an axis that is tilted at an angle of nearly 98° from the plane of its orbit. This means that, compared to all the other planets in the Solar System, Uranus is spinning on its side. During its 84-year orbit of the Sun, the North Pole of Uranus will have 42 years of continuous, sunny summer, while the South Pole has the same length of sunless winter, before they swap seasons. This odd axis may be the result of a catastrophic collision during the formation of the Solar System.

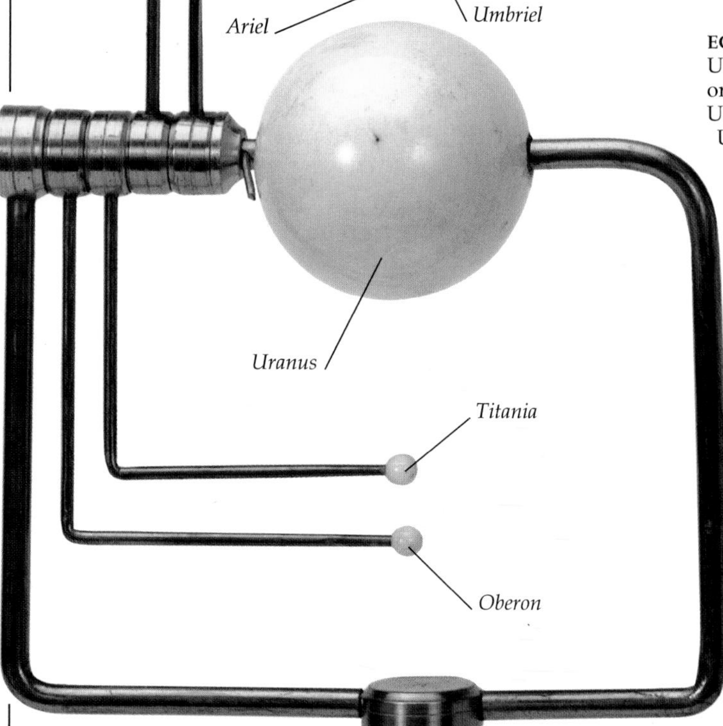

19TH-CENTURY MODEL
Because of the odd angle of Uranus's rotational axis, all its known satellites also revolve at right angles to this axis, around Uranus's equator. This fact is demonstrated by an early model, which shows the planet and four of its moons tilted at 98°. The planetarium (p. 36) dates from the 19th century when only four of the 15 moons had been discovered.

VIEW FROM SPACE
Uranus is one of the great gas giants of the Solar System, four times larger than the Earth. All that can be seen is the tops of its clouds. Unlike Jupiter and Neptune, however, the clouds of Uranus are largely featureless. These two images show true color (left) and false color, photographed from 5.7 million miles (9 million km) away.

AIRBORNE OBSERVATION OF URANUS
The covering of one celestial body by another is known as occultation. A team of scientists observed the occultation of Uranus in 1977 from NASA's Kuiper Airborne Observatory over the Indian Ocean. This was when the faint rings of Uranus were observed for the first time.

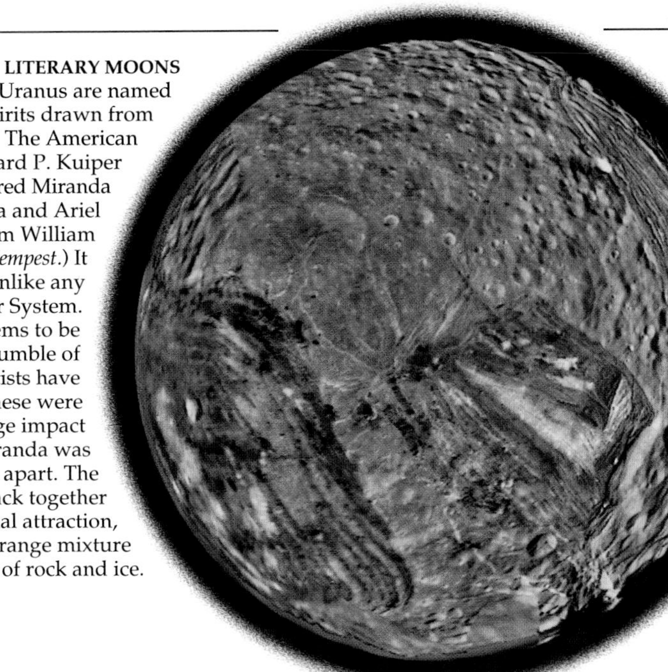

LITERARY MOONS
All the satellites of Uranus are named after sprites and spirits drawn from English literature. The American astronomer Gerard P. Kuiper (1905-1973) discovered Miranda in 1948. (Miranda and Ariel are characters from William Shakespeare's *The Tempest*.) It has a landscape unlike any other in the Solar System. Miranda seems to be composed of a jumble of large blocks. Scientists have suggested that these were caused by some huge impact during which Miranda was literally blown apart. The pieces drifted back together through gravitational attraction, forming this strange mixture of rock and ice.

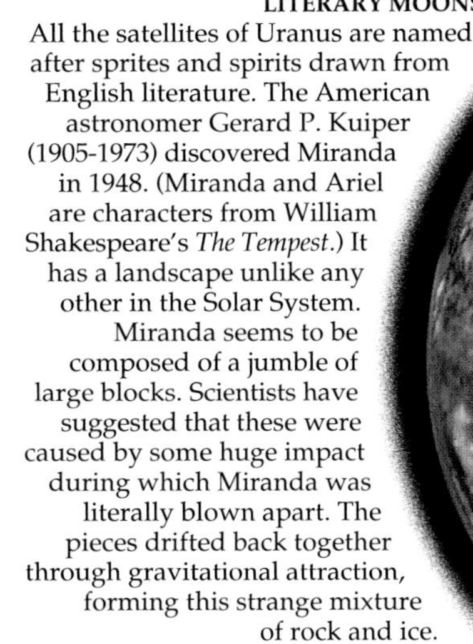

URANUS RING SYSTEM
While watching the occultation of Uranus in 1977, astronomers noticed that the faint star "blinked on and off" several times at the beginning and end of the occultation. They concluded that Uranus must have a series of faint rings that caused the star to "blink" by blocking off its light as it passed behind them. The *Voyager 2* flyby in 1986 uncovered two more rings. The rings of Uranus are thin and dark, made up of particles only about 39 in (1 metre) across. The broad bands of dust between each ring suggest that the rings are slowly eroding.

THE MOON TITANIA
William Herschel discovered Uranus's two largest moons in 1789, naming them Oberon and Titania, the fairy king and queen in William Shakespeare's *A Midsummer Night's Dream*. The English astronomer William Lassell (1799-1880) discovered Ariel and Umbriel in 1851. Miranda was discovered in 1948, and another eight moons were found during the 1986 *Voyager* flyby.

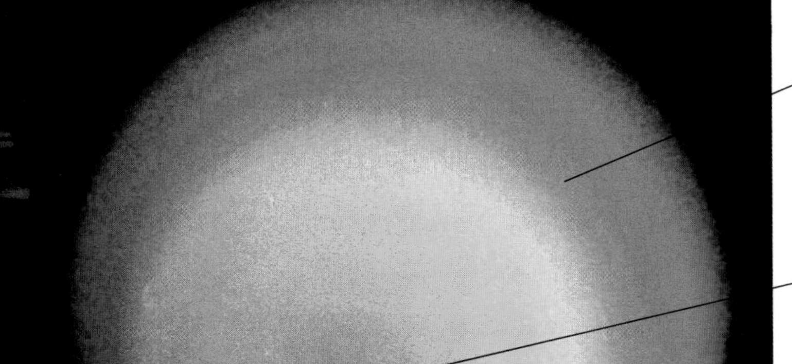

Color shows how haze affects the atmosphere

Polar region

Blemishes in processing

FACTS ABOUT URANUS

Hydrogen-rich atmosphere

Rocky core

Water, ammonia, and methane

- **Sidereal period** 84 Earth years
- **Temperature at cloud tops** –346°F
- **Rotational period** 17 hr 14 min
- **Mean distance from the Sun** 1,780 million miles/ 2,870 million km
- **Volume** (Earth = 1) 67
- **Mass** (Earth = 1) 14.5
- **Density** (water = 1) 1.29
- **Equatorial diameter** 31,760 miles/51,120 km
- **Number of satellites** 15

Neptune and Pluto

NEPTUNE AND PLUTO WERE DISCOVERED as the result of calculations rather than observations – although, in the case of Pluto, these calculations were later found to be pure coincidence. By the early 19th century, astronomers realized that Uranus has a slightly irregular orbit because "something" is pulling it out of line. This is known as a "perturbation" – the gravitational pull of one body takes another off its expected course. In 1845 a young English astronomer, John Couch Adams (1819-1892), announced that he had calculated the probable position of another planet beyond Uranus, then located against the stars of Aquarius. His findings were ignored. In June 1846 another young astronomer, the Frenchman Le Verrier, published his findings about the extra-Uranian planet. Astronomers all over the world were asked to search for the new planet. On September 25, 1846 Johann Galle (1812-1910), astronomer at the Berlin Observatory, wrote to Le Verrier confirming the location of Neptune.

URBAIN LE VERRIER (1811-1877)
Le Verrier was a teacher in chemistry and astronomy at the *Ecole polytechnique*. Having calculated the position of Neptune, Le Verrier relied on others to do the actual "looking" for the planet for him.

NEPTUNE'S RINGS
Neptune, like all the gas giants, has a series of rings encircling it. The rings were discovered when the planet passed in front of a star. Results of an occultation (p. 54) in July 1984 showed the typical "blinking on and off," indicating that Neptune's rings were blocking out the light of the distant star. There seem to be two main rings, with two faint inner rings. The inner ring is less than 9 miles (15 km) wide. The rings were confirmed by *Voyager 2* in 1989.

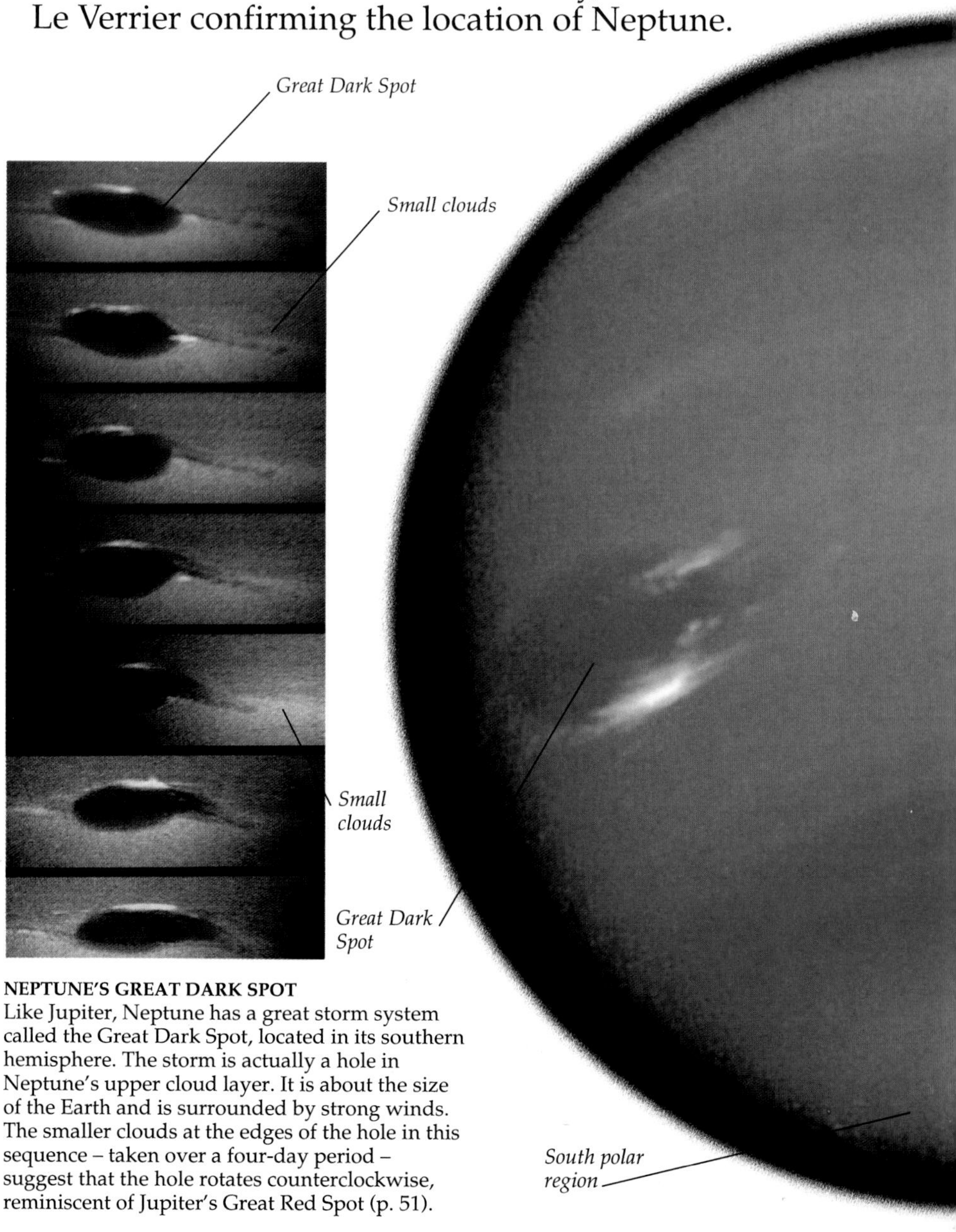

NEPTUNE'S GREAT DARK SPOT
Like Jupiter, Neptune has a great storm system called the Great Dark Spot, located in its southern hemisphere. The storm is actually a hole in Neptune's upper cloud layer. It is about the size of the Earth and is surrounded by strong winds. The smaller clouds at the edges of the hole in this sequence – taken over a four-day period – suggest that the hole rotates counterclockwise, reminiscent of Jupiter's Great Red Spot (p. 51).

FACTS ABOUT NEPTUNE

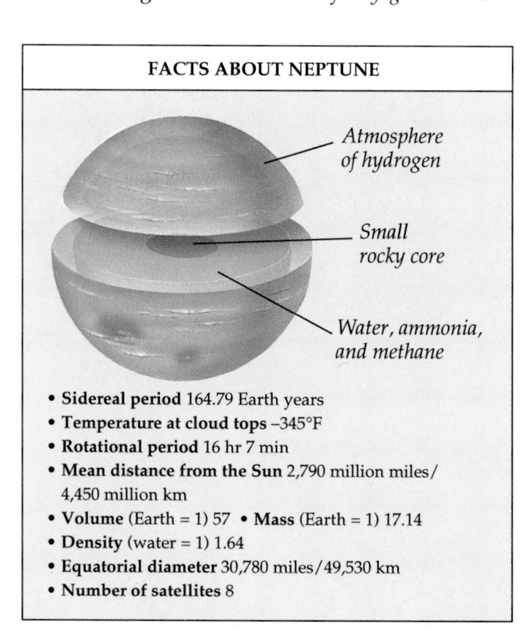

- **Sidereal period** 164.79 Earth years
- **Temperature at cloud tops** –345°F
- **Rotational period** 16 hr 7 min
- **Mean distance from the Sun** 2,790 million miles/ 4,450 million km
- **Volume** (Earth = 1) 57 • **Mass** (Earth = 1) 17.14
- **Density** (water = 1) 1.64
- **Equatorial diameter** 30,780 miles/49,530 km
- **Number of satellites** 8

Discovering Pluto

In 1930 an American astronomer, Clyde Tombaugh (1906-), discovered the planet that lay beyond Neptune. From night to night it shifted slowly against the background of the stars, showing that it was a long way away. An Oxford schoolgirl, Venetia Burney, suggested the name Pluto for the small, faint world, after the god of the underworld. Astronomers now know that Pluto is far too tiny to have any effect on Neptune's orbit. The calculations that led astronomers to look for Pluto were errors, and the discovery of the planet was due to Tombaugh's thorough work.

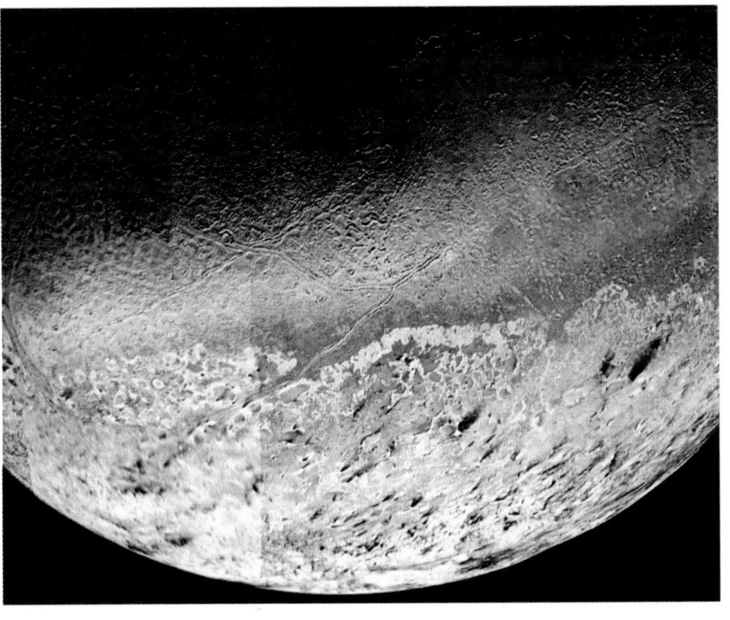

THE DISCOVERY OF TRITON
The moon Triton was discovered in 1846. It interests scientists for several reasons. It has a retrograde orbit around Neptune – that is, the moon moves in the opposite direction in which the planet rotates. It is also the coldest object in the Solar System with a temperature of –391°F (–235°C). Despite its low temperatures, Triton is a fascinating world. It has a pinkish surface, probably made of methane ice, which has repeatedly melted and refrozen. It has active geysers that spew nitrogen gas and darkened methane ice high into the thin atmosphere.

Sea-blue atmosphere

Ocean of water and gas

Smaller dark spot

CLOSE-UP OF NEPTUNE
This picture was taken by *Voyager 2* in 1989 after its 12-year voyage through the Solar System. It was 3.8 million miles (6 million km) away. *Voyager* went on to photograph the largest moon, Triton, and to reveal a further six moons orbiting the planet. Neptune has a beautiful, sea-blue atmosphere, composed mainly of hydrogen and a little helium and methane. This covers a huge internal ocean of warm water and gases – appropriate for a planet named after the god of the sea. (Many French astronomers had wanted the new planet to be named "Le Verrier," in honor of its discoverer.) *Voyager 2* discovered several storm systems on Neptune, as well as beautiful white clouds high in the atmosphere.

PERCIVAL LOWELL (1855-1916)
Lowell was born into a wealthy Boston family and was educated at Harvard. His interest in astronomy was stimulated by the reports of dark lines on Mars, and until his death he was a firm believer in intelligent life on Mars (p. 48). Later Lowell became embroiled in a race to discover the planet beyond Neptune. Because he was looking for a planet that he believed is seven times heavier than the Earth, Lowell missed discovering Pluto, although the planet appeared on several of his photographs of the skies.

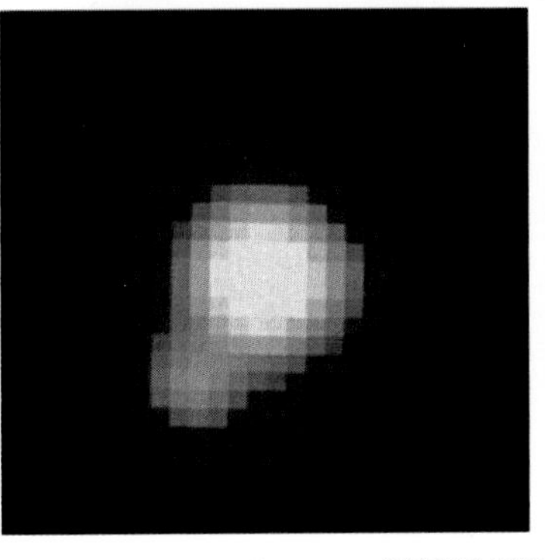

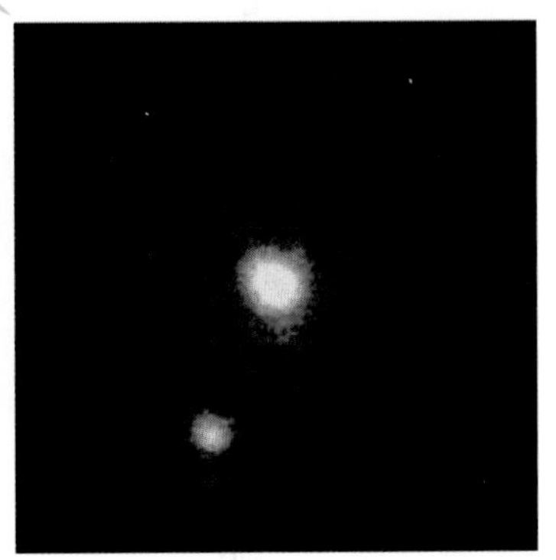

PLUTO AND CHARON
The distance between Pluto and its moon, Charon, is only 12,240 miles (19,700 km). Charon was discovered in 1978 by a study of images of Pluto that looked suspiciously elongated. The clear image on the right was taken by the Hubble telescope (p. 7), which allows better resolution than anything photographed from the Earth (left).

FACTS ABOUT PLUTO

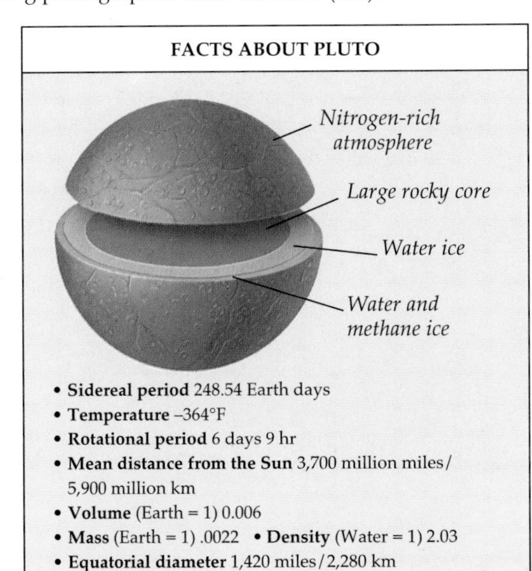

- **Sidereal period** 248.54 Earth days
- **Temperature** –364°F
- **Rotational period** 6 days 9 hr
- **Mean distance from the Sun** 3,700 million miles/ 5,900 million km
- **Volume** (Earth = 1) 0.006
- **Mass** (Earth = 1) .0022 • **Density** (Water = 1) 2.03
- **Equatorial diameter** 1,420 miles/2,280 km
- **Number of satellites** 1

Travelers in space

NOT ALL MATTER IN THE SOLAR SYSTEM has been brought together to form the Sun and the planets. Clumps of rock and ice travel through space, often in highly elliptical orbits that carry them toward the Sun from the far reaches of the Solar System. Comets are icy planetary bodies that take their name from the Greek description of them as *aster kometes*, or "long-haired stars." Asteroids are bits of rock that have never managed to come together as planets. Scientists are not sure why, but they think that Jupiter's powerful gravity flung the nearby debris into various eccentric orbits, causing some of the larger asteroids to crash together and break up. A meteor is a piece of space rock – usually a small piece of a comet – that enters the Earth's atmosphere. As it falls, it begins to burn up, producing spectacular fireworks. A meteor that survives long enough to hit the Earth – usually a stray fragment from the asteroid belt – is called a meteorite.

PREDICTING COMETS
Going through astronomical records in 1705, Edmond Halley (1656-1743) noticed that three similar descriptions of a comet had been recorded at intervals of 76 years. Halley used Newton's recently developed theories of gravity and planetary motion (p. 21) to deduce that these three comets might be the same one returning to the Earth at regular intervals, because it was traveling through the Solar System in an elliptical orbit (p. 13). He predicted that the comet would appear again in 1758, but he did not live to see the return of the comet that bears his name.

COMET'S TAIL
Comets generally have elongated orbits. They can be seen by the light they reflect. As they get closer to the Sun's heat, their surface starts to evaporate and a huge tail of steamy gas is given off. This tail always points away from the Sun because the dust and gas particles are pushed by solar wind and radiation pressure.

Nucleus

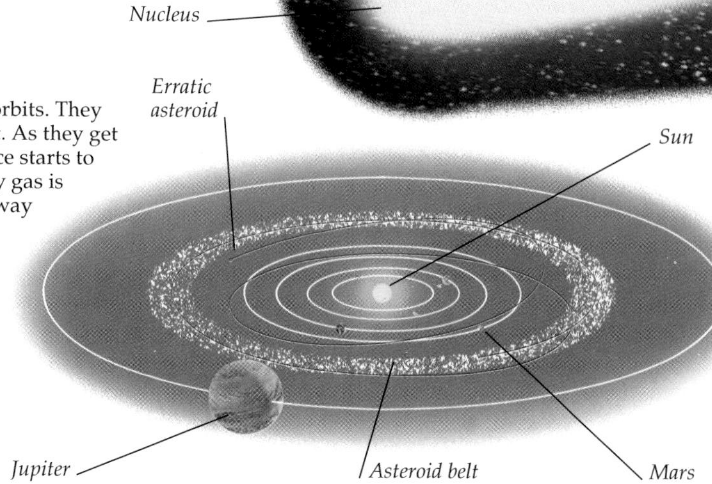

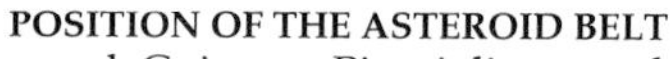

POSITION OF THE ASTEROID BELT
Since the Sicilian monk Guiseppe Piazzi discovered the first asteroid in January 1801, more than 4,000 asteroids have been found and named. Most of them travel in a belt between Mars and Jupiter, but Jupiter's great gravitational influence has caused some asteroids to swing out into erratic orbits.

KORONIS FAMILY OF ASTEROIDS
This photograph of the asteroid Ida was taken by the *Galileo* spacecraft in 1993 as the space probe traveled to Jupiter. The cratered surface probably resulted from collisions with smaller asteroids. Ida is 32 miles (52 km) long.

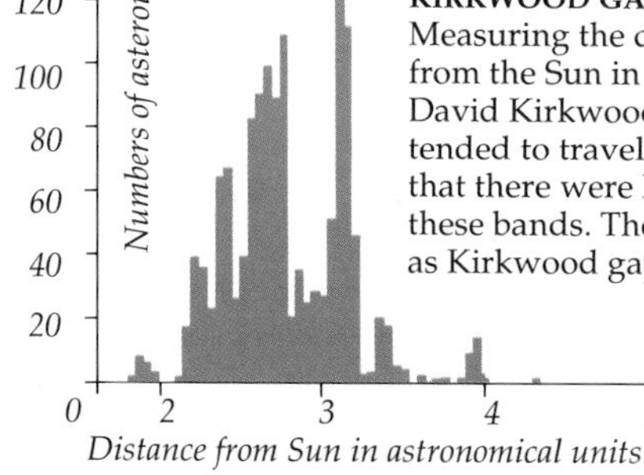

KIRKWOOD GAPS
Measuring the distances of the known asteroids from the Sun in 1866, the American astronomer David Kirkwood (1814-1895) noticed that they tended to travel in loosely formed bands and that there were large, peculiar gaps between these bands. The gaps, which are now known as Kirkwood gaps, are due to recurring "bumps" from Jupiter's gravitational field. Asteroids can be catapulted into the inner Solar System by Jupiter's gravity.

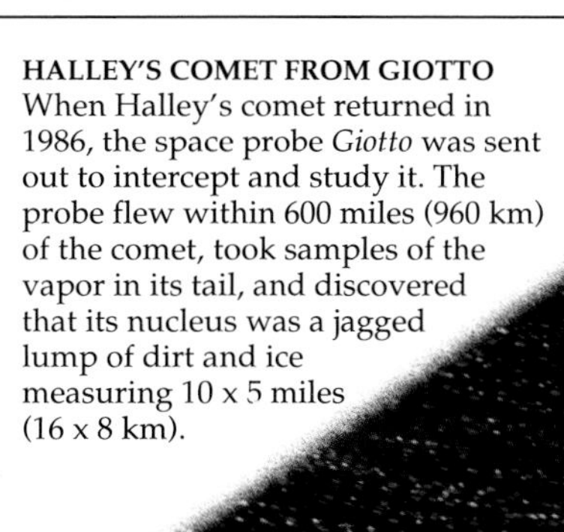

HALLEY'S COMET FROM GIOTTO
When Halley's comet returned in 1986, the space probe *Giotto* was sent out to intercept and study it. The probe flew within 600 miles (960 km) of the comet, took samples of the vapor in its tail, and discovered that its nucleus was a jagged lump of dirt and ice measuring 10 x 5 miles (16 x 8 km).

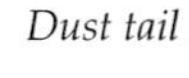

Plasma (gas) tail

Meteorites

It was not until 1803 that the scientific community accepted that meteorites did, indeed, fall from space. Over 95 percent of all the meteorites recovered are stone meteorites. Meteorites are divided into three types with names that describe the mix of elements found within each specimen. Stony meteorites look like stones but usually have a fused crust caused by intense heating as the meteorite passes through the Earth's atmosphere. Iron meteorites contain nickel iron crystals, and stony iron meteorites are part stone, part iron.

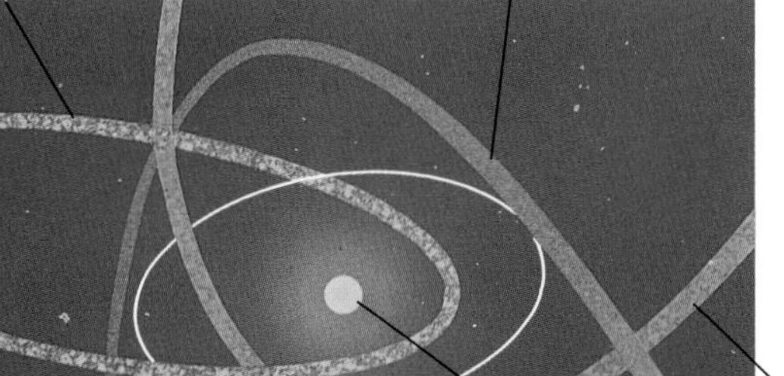

METEOR SHOWERS
When the Earth's orbit cuts through a stream of meteors, the meteoritic material seems to radiate out from one point in the sky, creating a meteor shower. The showers are given names, such as Geminid, that derive from the constellations in the sky from which they seem to be coming.

Tektite

MOLTEN DROPLET
Tektites are small, round, glassy objects that are usually the size of marbles. They are most often found on the Earth in great numbers, all together. When a blazing meteorite hits a sandstone region, the heat temporarily melts some of the metals in the Earth's soil. These molten droplets harden to form tektites.

Murchison meteorite

METEOR IN AUSTRALIA
This meteorite (left) fell near the Murchison River in Western Australia in 1969. It is made up of carbon and water. The carbon comes from chemical reactions and not from once-living organisms like those carbon compounds that are found on the Earth, such as coal.

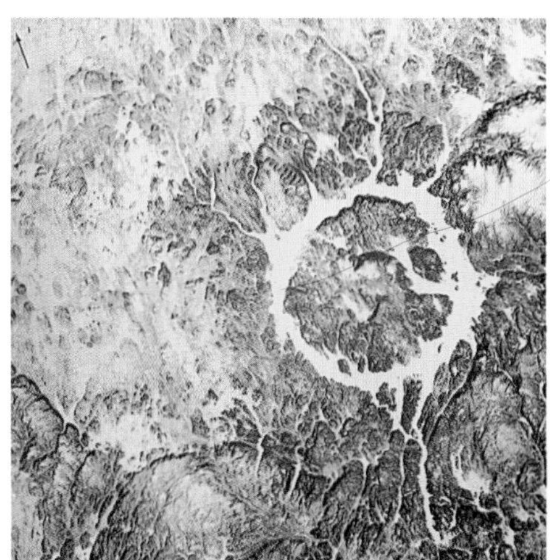

ICY CRATER
The Earth bears many scars from large meteorites, but the effects of erosion and vegetation cover over some of the spectacular craters. This space view shows an ice-covered crater near Quebec in Canada. It is now a 41-mile (66-km)-wide reservoir used for hydroelectric power.

The birth and death of stars

APART FROM THE SUN, the closest star to the Earth is Proxima Centauri, which is 4.2 light-years or 25 trillion miles (40 trillion km) away. A light-year is the distance that light or other electromagnetic radiation (p. 32) travels in a year. Stars are luminous, gaseous bodies that generate energy by means of nuclear fusion in their cores (pp. 38-39). As a star ages, it uses up its fuel. The core shrinks under its own weight while the nuclear "burning" continues. The shrinkage heats up the core, making the outer layers of the star expand and cool. The star becomes a "red giant." The more massive stars will continue to fuse all their lighter elements until they reach iron. As the remains of the star's atmosphere escapes, it leaves the core exposed as a "white dwarf." When a star tries to fuse iron, there is a massive explosion and the star becomes a "supernova." After the explosion, the star's core may survive as a pulsar or a "black hole" (p. 62).

CATALOG OF STARS
The French astronomer Charles Messier (1730-1817) produced a catalog of around 100 of the brightest nebulae (clouds of dust and gas) in 1784. Each object was numbered and given an "M-" prefix. For example, the Orion nebula is the 42nd object in Messier's list. Astronomers usually refer to this nebula as M42.

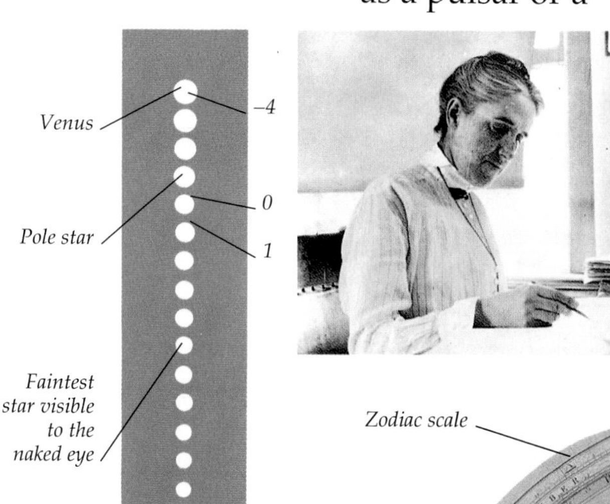

HENRIETTA LEAVITT (1868-1921)
In 1912 the American astronomer Henrietta Leavitt was studying Cepheid variable stars. These are a large group of bright yellow giant and supergiant stars named after their prototype in the constellation of Cepheus. Variable stars are stars that do not have fixed brightness. Leavitt discovered that the brighter stars had longer periods of light variation. This variation can be used to determine stellar distances beyond 100 light-years.

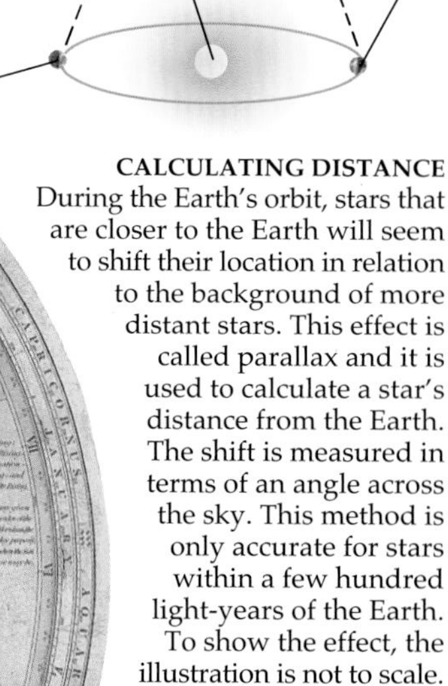

CALCULATING DISTANCE
During the Earth's orbit, stars that are closer to the Earth will seem to shift their location in relation to the background of more distant stars. This effect is called parallax and it is used to calculate a star's distance from the Earth. The shift is measured in terms of an angle across the sky. This method is only accurate for stars within a few hundred light-years of the Earth. To show the effect, the illustration is not to scale.

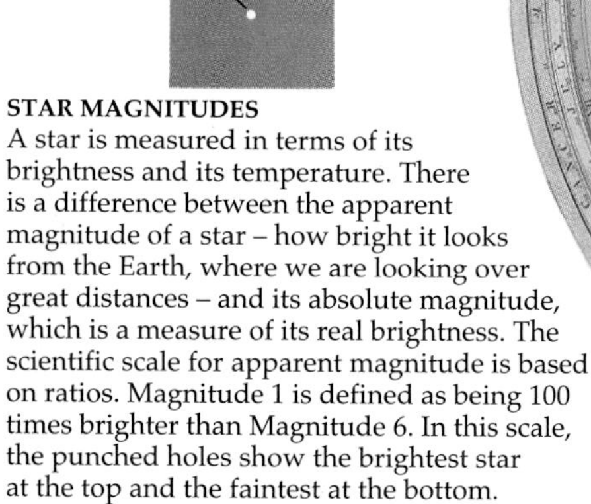

STAR MAGNITUDES
A star is measured in terms of its brightness and its temperature. There is a difference between the apparent magnitude of a star – how bright it looks from the Earth, where we are looking over great distances – and its absolute magnitude, which is a measure of its real brightness. The scientific scale for apparent magnitude is based on ratios. Magnitude 1 is defined as being 100 times brighter than Magnitude 6. In this scale, the punched holes show the brightest star at the top and the faintest at the bottom.

A MAP OF THE STARS
Since ancient times, astronomers have had difficulties in being able to translate what is essentially a three-dimensional science into the medium of two dimensions. One solution was the planisphere, or "flattened sphere," in which the whole of the heavens was flattened out with the Pole Star at the center of the chart.

STUDYING THE STARS

The British astronomer Williams Huggins (1824-1910) was one of the first to use spectroscopy for astronomical purposes (pp. 30-31). He was also the first astronomer to connect the Doppler effect (which relates to how sound travels) with stellar red shift (p. 22). In 1868 he noticed that the spectrum of the bright star Sirius has a slight shift toward the red end of the spectrum. He correctly deduced that this effect is due to that star traveling away from the Earth. He also discovered that nebulae are made up of luminous gases.

NOVAE AND SUPERNOVAE

A nova was originally believed to be a newborn star. It is now known that novae are not new stars, but close double stars that flare up as one star loads its matter on to another. A supernova is much more violent, and takes place when a massive star collapses at the end of its life. Its core may, however, survive to become a small pulsating star (pulsar), or even a black hole. The gas blown off by the explosion forms an expanding shell. The gas from the supernova triggers the birth of new stars. These pictures show the appearance before (left) and after (right) of supernova 1987A over two weeks in 1987.

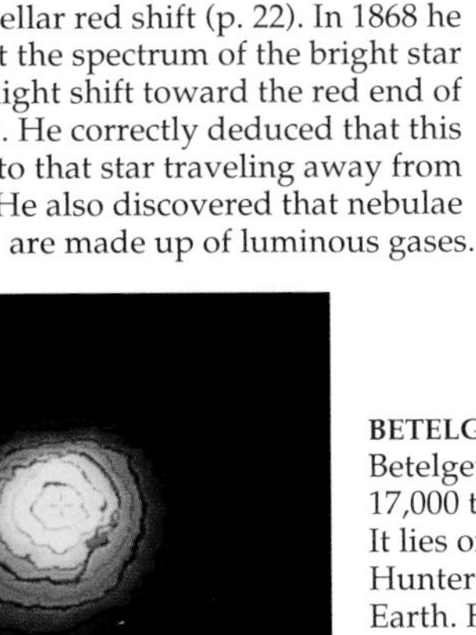

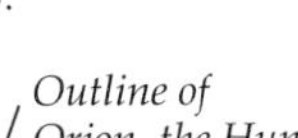

BETELGEUSE

Betelgeuse is a variable star that is 17,000 times brighter than our Sun. It lies on the shoulder of Orion the Hunter, 650 light-years from the Earth. False colour has been used here to highlight the gases around the star. Astronomers believe that it will "die" in a supernova explosion (above right).

Outline of Orion, the Hunter

Bellatrix

Rigel

THE CONSTELLATION ORION

A constellation is a group of stars that appear to be close to each other in the sky, but that are usually spread out in three-dimensional space. Orion the Hunter's stars include the bright Betelgeuse and Rigel.

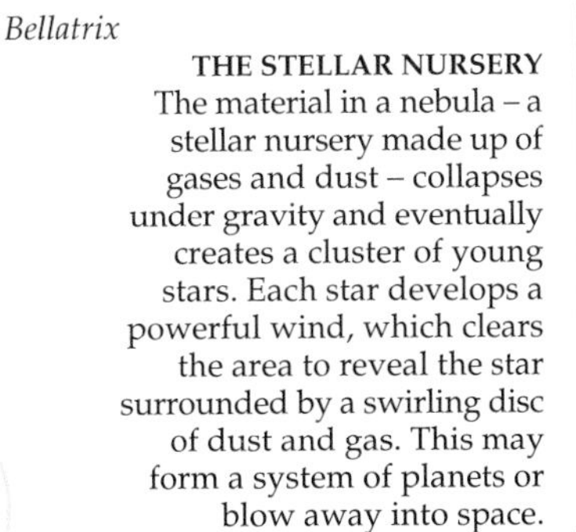

THE STELLAR NURSERY

The material in a nebula – a stellar nursery made up of gases and dust – collapses under gravity and eventually creates a cluster of young stars. Each star develops a powerful wind, which clears the area to reveal the star surrounded by a swirling disc of dust and gas. This may form a system of planets or blow away into space.

ORION NEBULA M42

Stars have a definite life cycle that begins in a mass of gas that turns into stars. This "nebula" glows with color because of the cluster of hot, young stars within it. This is part of the Great Nebula in Orion.

Our galaxy and beyond

MOST GALAXIES WERE FORMED about three billion years after the Universe was born. As it contracted under its own gravity, each galaxy took on a certain shape. A typical galaxy contains about 100 billion stars and is around 100,000 light-years in diameter. Edwin Hubble was the first astronomer to study these distant star systems systematically. While observing the Andromeda Galaxy in 1923, he was able to measure the brightness of some of its stars. From this information, he deduced that they are 2.25 million light-years away (p. 60). After studying the different red shifts of the galaxies (p. 23), Hubble proposed that the galaxies are moving away from our galaxy at speeds proportional to their distances from us. His law shows that the Universe is expanding.

EDWIN HUBBLE (1889-1953)
In 1923 the American astronomer Edwin Hubble studied the outer regions of what appeared to be a nebula (p. 61) in the constellation of Andromeda. With the high-powered 100-in (254-cm) telescope at Mount Wilson, he was able to see that the "nebulous" part of the body is composed of stars, some of which were bright, variable stars called Cepheids (pp. 60-61). Hubble realized that for these intrinsically bright stars to appear so dim, they must be extremely far away from the Earth. His research helped astronomers begin to understand the immense size of the Universe.

THE MILKY WAY
From the Earth, the Milky Way appears particularly dense near the constellation of Sagittarius because this is the direction of the galaxy's center. Although optical telescopes cannot penetrate the galactic center because there is too much interstellar dust in the way, radio and infrared telescopes can.

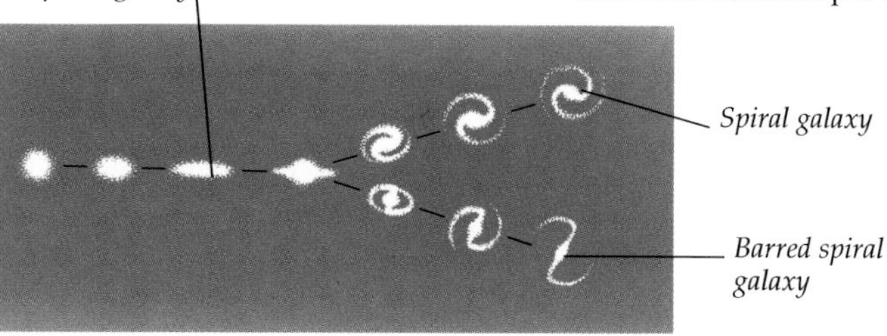

CLASSIFYING GALAXIES
Hubble devised a classification of galaxies according to shape. Elliptical galaxies were subdivided by how flat they appeared. He classified spiral and barred spiral galaxies (where the arms spring from a central bar) according to the tightness of their arms.

The Milky Way photographed from Chile with a wide-angle lens

Observatory building

WHIRLPOOL GALAXY
The Whirlpool Galaxy is a typical spiral galaxy, approximately 14 million light-years away. It can be found among the faint group of stars Canes Venatici, at the end of the tail of the constellation of Ursa Major, or the Great Bear. It was one of the nebulae drawn by Lord Rosse (p. 26) in the 19th century.

THE SHAPE OF OUR GALAXY
The Milky Way, seen edge-on (top), has an oval central bulge surrounded by a very thin disc containing the spiral arms. It is approximately 100,000 light-years in diameter and about 15,000 light-years thick at its center. Our Sun is located about 30,000 light-years away from the center. The Milky Way looks like a band in our skies because we see it from "inside" – its disc is all around us. Viewed from above (bottom), it is a typical spiral galaxy with the Sun situated on one of the arms, known as the Orion arm.

ANDROMEDA GALAXY
The Andromeda Galaxy is a spiral galaxy, shaped like our Milky Way, but it has nearly half as much mass again. It is the most distant object that is visible to the unaided eye. It has two small elliptical companion galaxies.

What is cosmology?

Cosmology is the name given to the branch of astronomy that studies the origin and evolution of the Universe. It is an ancient study, but in the 20th century the theory of relativity, advances in particle physics and theoretical physics, and the discoveries about the expanding Universe have given cosmology a more scientific basis and approach.

ALBERT EINSTEIN (1879-1955)
In proposing that mass is a form of energy, the great German-American scientist Albert Einstein redefined the laws of physics dominant since Newton's time (p. 22). The fact that gravitation could affect the shape of space and the passage of time meant that scientists were finally provided with the tools to understand the birth and death of the stars, especially the phenomenon of the black hole.

BLACK HOLES
A supernova (p. 61) can leave behind a black hole – an object that is so dense and so collapsed that even light cannot escape from it. Although black holes can be detected when gas spirals into them, because the gases emit massive quantities of X-rays as they are heated, they are otherwise very hard to find. Sometimes they act as "gravitational lenses," distorting background starlight.

Index

Acknowledgments

Dorling Kindersley would like to thank:
Maria Blyzinsky for her invaluable assistance in helping with the objects at the Old Royal Observatory; Peter Robinson and Artemi Kyriacou for modelling; Peter Griffiths for making the models; Jack Challoner for advice; Frances Halpin for assistance with the laboratory experiments; Paul Lamb, Helen Diplock, and Neville Graham for helping with the design of the book; Anthony Wilson for reading the text; Harris City technology College and The Royal Russell School for the loan of laboratory equipment; the Colour Company and the Roger Morris Partnership for retouching work; lenses supplied by Carl Lingard Telescopes, 89 Falcon Crescent, Clifton, Swinton, Manchester.

Illustrations Janos Marffy, Nick Hall, John Woodcock and Eugene Fleury

Photography Colin Keates, Harry Taylor, Christi Graham, Chas Howson, James Stevenson and Dave King.
Index Jane Parker

Picture credits
t=top b=bottom c=center l=left r=right

American Institute of Physics, Emilio Segrè Visual Archives/Bell Telephone Laboratories 32cl;/W. F. Meggers Collection 32tl; /Research Corporation 32bl;/Shapley Collection 60cl.
Ancient Art and Architecture Collection 9tl; 9cr; 20tl.
Anglo-Australian Telescope Board/D. Malin 61tr.
Archive für Kunst und Geschichte, Berlin 19tl;/National Maritime Museum 46c.
Associated Press 8tl.
The Bridgeman Art Library 28tl;/Lambeth Palace Library, London 17tr.
The Observatories of the Carnegie Institution of Washington 39cr.
Jean-Loup Charmet 60tl.
Bruce Coleman Ltd. 43c.
ET Archive 9tr.
Mary Evans Picture Library 6tl; 14bl; 18cb; 42tl; 61tcl.
Robert Harding/R. Frerck 32bc; /C. Rennie 10br. Hulton Deutsch 20c.
Henry E. Huntington Library and Art Gallery 62tl. Images Colour Library 7tl; 7tr; 7c; 16tl; 16cl; 18cl. Image Select 19br; 21tc; 23tl; 26cl; 28cr. JPL 3cr; 37cr; 44bl; 48br; 49tl; 49c; 49cr; 50tr; 51crb; 52cl; 52br; 53cl; 55tr; 56cl; 63tr. Lowell Observatory 48tr. Magnum/E. Lessing 19tr.
Mansell Collection 49tr.
NASA 3cr; 8cl; 35cr; 35bl; 38/39bc; 41cr; 44cl; 46br; 47b; 49br; 54/55bc; 56/57bc; 57tl; /JPL 34bc; 39tc; 50br; 51t; 51bc; 53r; 55c; 55cr; 56cb; 61cr.
National Geophysical Data Centre/ NOAA 27cl.
National Maritime Museum Picture Library 6cl; 10cr; 15tl; 15tr; 25tl; 27br; 29tr; 39c; 54tl. Novosti (London)28br; 35cl; 47tr;/Tass 25clb.
Popperfoto 47tl; 67crb. Rex Features Ltd. 34clb. Scala/Biblioteca Nationale 20cr.
Science Photo Library jacket c; 18tl; 25bc; 31tl; 31bl; 48c; 53tl; 56tl; 57cr;/ Dr. J. Burgess 26tl;/Jean-Loup Charmet 11tl; 37tl; /F. Espenak 46tr;/European Space Agency 35tl; 42/43bc;/Jodrell Bank 33tr;/Dr. M. J. Ledlow 33tl;/F. D. Miller 31c;/NASA 7crb; 20bc; 20bcr; 20br; 34cr; 41crb; 44/45bc; 49tc; 57cr; 58bl; 59br; /NOAO 61cl;/Novosti 47cr;/P. Plailly 38cl;/Physics Dept. Imperial College 30crb; 30br;/Dr. M. Read 30tl;/Royal Observatory, Edinburgh 27tl; 59tr; /J. Sandford 41t; 61bl;/R. Ressemeyer, Starlight 12tl; 25crb; 27tr; 33cr; 55tl; 62/63c;/U.S. Geological Survey 37br; 48l.
Tony Stone Images 42cl. Roger Viollet/Boyes 38tl. ZEFA UK 6/7bc; 8bl; 33b; 61br; 62bl; /G. Heil 26bc.

With the exception of the items listed above, the object from the British Museum on page 8c, from the Science Museum on pages 21b, 22 cl, and from the Natural History Museum on page 43tl, the objects on pages 1, 2t, 2c, 2b, 3t, 3l, 3b, 3tr, 4, 11b, 12b, 14c, 14bc, 14r, 15tl, 15b, 16bl, 16br, 17bl, 17br, 20bl, 24bc, 25tr, 28cl, 28c, 28bl, 29tl, 29c, 29b, 31b, 36b, 38b, 40cl, 40bl, 40/4b, 42bl, 52bl, 54bl, 58tr, 60b, are all in the collection of the Old Royal Observatory, Greenwich.

1 BIRD
2 ROCKS & MINERALS
3 SKELETON
4 ARMS & ARMOR
5 TREE
6 POND & RIVER
7 BUTTERFLY & MOTH
8 SPORTS
9 SHELL
10 EARLY HUMANS
11 MAMMAL
12 MUSIC
13 DINOSAUR
14 PLANT
15 SEASHORE
16 FLAG
17 INSECT
18 MONEY
19 FOSSIL
20 FISH
21 CAR
22 FLYING MACHINE
23 ANCIENT EGYPT
24 ANCIENT ROME
25 CRYSTAL & GEM
26 REPTILE
27 INVENTION
28 WEATHER
29 CAT
30 BIBLE LANDS
31 EXPLORER
32 DOG
33 HORSE
34 FILM
35 COSTUME
36 BOAT
37 ANCIENT GREECE
38 VOLCANO & EARTHQUAKE
39 TRAIN
40 SHARK
41 AMPHIBIAN
42 ELEPHANT
43 KNIGHT
44 MUMMY
45 COWBOY
46 WHALE
47 AZTEC, INCA & MAYA
48 BOOK
49 CASTLE
50 VIKING
51 DESERT
52 PREHISTORIC LIFE
53 PYRAMID
54 JUNGLE
55 ANCIENT CHINA
56 ARCHEOLOGY
57 ARCTIC & ANTARCTIC
58 BUILDING
59 PIRATE
60 NORTH AMERICAN INDIAN
61 AFRICA
62 OCEAN
63 BATTLE
64 GORILLA, MONKEY & APE
65 MEDIEVAL LIFE
66 FARM
67 SPY
68 RELIGION
69 EAGLE & BIRDS OF PREY
70 WITCHES & MAGIC-MAKERS
71 SPACE EXPLORATION
72 SHIPWRECK